The Patti Playpal Family

A Guide to Companion Dolls of the 1960s

Carla Marie Cross

4880 Lower Valley Road, Atglen, PA 19310 USA

Published by Schiffer Publishing Ltd.
4880 Lower Valley Road
Atglen, PA 19310
Phone: (610) 593-1777; Fax: (610) 593-2002
E-mail: Schifferbk@aol.com
Please visit our web site catalog at
www.schifferbooks.com

This book may be purchased from the publisher.
Include $3.95 for shipping.
Please try your bookstore first.
We are interested in hearing from authors
with book ideas on related subjects.
You may write for a free catalog.

In Europe, Schiffer books are distributed by
Bushwood Books
6 Marksbury Ave.
Kew Gardens
Surrey TW9 4JF England
Phone: 44 (0) 208 392-8585; Fax: 44 (0) 208 392-9876
E-mail: Bushwd@aol.com
Free postage in the U.K., Europe; air mail at cost.
Try your bookstore first

Copyright © 2000 by Carla Marie Cross
Library of Congress Catalog Card Number: 00-101014

All rights reserved. No part of this work may be reproduced or used in any form or by any means—graphic, electronic, or mechanical, including photocopying or information storage and retrieval systems—without written permission from the copyright holder.
"Schiffer," "Schiffer Publishing Ltd. & Design," and the "Design of pen and ink well" are registered trademarks of Schiffer Publishing Ltd.

Book Design by Anne Davidsen
Type set in Kids / Humanist

ISBN: 0-7643-1146-8

Printed in China
1 2 3 4

Contents

Dedication
4

Acknowledgments
5

Chapter One
What A Big Girl You Are!
Information on Playpal and companion dolls, Neil Estern, sculptor
6

Chapter Two
Patty Playpal
Original outfits, hair colors and styles, 1981 reissue, 1986 reissue
13

Chapter Three
Playpal Family Members
Pattite, Peter, Penny, Suzy, Twins Bonnie and Johnny
58

Chapter Four
The 36" Shirley Temple Dolls
76

Chapter Five
Playsets, Original TV & Catalog Ads
88

Chapter Six
Portrait Gallery
90

Chapter Seven
Other Large Dolls of the Era
95

Dedication

This book is in appreciation of all Playpal dolls –
Not just dolls, but friends to the children…

"(S)he lies in the laurels,
      (S)he runs on the grass;
(S)he sings when you
      tinkle the musical glass;
When you are happy
      And cannot tell why,
The friend of the children
      Is sure to be by!

From the poem "The Unseen Playmate" by Robert Louis Stevenson. Child pictured is author's daughter, Jonnica

Acknowledgments

A heartfelt THANK YOU to all photo contributors, and especially to John Sonnier, John Medeiros, Murray Hilliard, Kathy Hosteller, Kathy Ebey, Laura Brink, Tara Wood, and Lori Gabel.
AND to my photographer's assistants: Tiffin, Dustin, Lacelynn, Jamison, Bria, Jonnica and Mckaryn Cross.
All photography by the author, unless otherwise indicated.

Chapter One
What a Big Girl You Are!

    The affection of a young girl for her dolls could rival the very best of romantic fairy tales. Dolls touch the heart and develop compassion in young children. There is tremendous worth in a child dressing, caressing, rubbing noses with and generally relating to a doll. The early dawnings of maternal/paternal instincts are formed here, and the treasure trove of childhood memories they create are priceless. As a class, actual play dolls may not shine as brightly in the eyes of some collectors. But in reality, they amass far more real value in their "lifetimes" because of what they end up giving to children. 1960s advertising executives quickly picked up the American girl's love affair with dolls, and as a result, dolls of that era became the real celebrities of television ads and Christmas catalogs. Trains and bikes had their place, of course, but almost never occupied as many pages or as much air time.

    In Victorian times, many children had only one doll, which was thoroughly cherished. Playpal dolls are special for this same reason. At the time the Playpal dolls were made, it was not unusual for a child to have a number of 11 1/2 inch fashion dolls, or small drink and wet babies, but, because of their size and cost, most children had only one Playpal doll.

    Dolls, in the real sense of the word, have not changed much over the years, except for the material used for their manufacture. Baby dolls have been around since the early 1800s and dolls meant to look like adults seem to be as old as recorded history.

    Ideal's Playpal dolls, however, stand out in this very long line for two reasons: 1) Their size - roughly the size of the child playing with them - is a unique feature in the history of dolls.

> ## INTERESTING FACT
>
> "Playpal" was spelled Play Pal - as two separate words in 1959 - the first year of production. In 1960, the spelling became "Playpal" and stayed that way until 1987, when Ideal issued Talking Patty Play Pal.

2) They are distinctive works of art, both in sculpture and in design and can be appreciated as more than just a toy. For example, an individual with little or no interest in dolls can walk through a doll show or doll museum completely unimpressed until they fall upon a Playpal doll. "Companion" or life-size dolls, as they are called, are quite impressive to someone caught off guard, because, for a brief moment, it's like encountering a real person.

What constitutes a companion doll? I think the keyword is not size, but "life-size." A baby doll, for example, that is 23 inches or larger, a toddler doll that is 28 inches or larger, or a doll representing a child that is 35 inches or larger, would be considered a companion doll, because they are the size of their real life counterparts. On the other hand, a 30 inch lady doll, even though she is large, would not be considered a companion doll.

Some of the fun of owning companion-sized dolls is that they can be dressed in clothing from the children's department of a store, so the variety is almost endless. Patti Playpal wears about a size 3T, Peter a size 4T, Penny about a size 2T, Suzy about a size 12-18 months, and Bonnie and Johnny about a size 3-6 months, all of these depending on individual brand names. These dolls can be used to display a family heirloom such as a christening gown or a special dress or coat from someone's childhood. Some collectors will dress their companion dolls up in Christmas and Easter fancy dress, or in Halloween costumes, and set them outside to intrigue the neighborhood children.

Collecting this line of dolls can be a real adventure. Just when you think you've seen everything "Playpal," something pops up, like a new hair color or style, or a variation of one of the more common outfits. The pursuit of Playpal original outfits can have all the earmarks of a Nancy Drew investigation. Some, but not all, of the Playpal outfits are tagged. When an original outfit is not tagged, a collector has to search through old store catalogs, or request old photos of the original owner actually receiving the doll on Christmas morning or for a birthday to see if the untagged dress is indeed authentic. Patti herself has so many original dresses, I'm never surprised when another one is discovered.

Where to Find Playpals

Authentic Playpals, marked Ideal, are not generally found at flea markets or rummage sales, the way smaller dolls are. Older dolls that surface at rummage-type sales are often there because they have been stored away for a number of years and someone finally decided to do a general housecleaning. Playpal dolls, for the most part, were just too big to be stored for very long. When the children in the family had outgrown them, they weren't often tucked away in a drawer or closet for the grandchildren. They, like the boxes they came in, were large and cumbersome, and somewhat inconvenient to keep around. The original boxes were thrown out and the dolls were often given to second hand stores or church sales at a time when they weren't tremendously collectible. Many of the very early Playpal collectors can remember picking up these dolls in the 70s for just a few dollars, and, of course, have hung on to them. In today's market, therefore, antique malls, doll shows, trade papers such as Collectors United and Master Collector, and on-line auction sites are your best bet!

Patti, Penny and Peter Playpal are, in general, found in better condition than most play dolls of that era. Because of their size, they weren't knocked around as much, or dragged outside to the sandbox and left there. The biggest concern for Pattis is an occasional haircut. Often, limbs will have faded to a distinctly different color than the torso, because of the different type of vinyl used. The same kind of color difference can be seen between the hands and the arms of the early "twist wrist" Pattis. This color difference does not depreciate the value of a Patti by very much, as it is seen as commonplace by collectors, and they have grown used to it. Once in a while, a Patti will be found with a "sticky eye" (one sleep eye that doesn't close as well as the other when the doll is laid down). This, also, does not depreciate the doll's value by much if at all - 5% at the very most.

What to Look For

PATTI. If her body is strung, see if she can stand on her own two feet. If the stringing has loosened over the years, or if she has been poorly restrung, her legs will often slip out of their sockets as they try to support her in a standing position.

SMELL. Because of Patti, Penny, and Peter's large size the moms that did want to keep the dolls could not always store them in the living area of the house. The Playpal dolls that were not given to the Salvation Army, etc., were often stored in basements. As these few dolls find their way to the collector's market after 30 or more years of basement life, the damp and musty scent they have acquired can be overwhelming. If you are purchasing a Playpal by mail order or on line from an individual you are not familiar with, ASK ABOUT SMELL.

HAIRCUTS: A factory haircut and one administered by a well-intentioned 4 year old will have distinctly different looks to them. Carefully examine the ends of the doll's hair. Patti Playpals with the longer hairstyles often have some variation and unevenness to the ends of the hair, but will look monumentally better than a child's efforts. If you feel that the hair has been cut, remember that this devalues a doll tremendously.

RARITIES: Certain Playpal hair colors and styles are very hard to find, and consequently, more coveted by collectors. Certain outfits and playsets, as well, are rare and more pricey. Look through the rarities in this book to help you identify and value these dolls and items.

General Information

Ideal knew that they had some design details to work through in producing a life-size doll. Under normal manufacturing practices, using a vinyl as solid as some of their smaller Ideal counterparts, these dolls would have been far too heavy for a child to handle comfortably. The Playpal dolls, therefore, were made of a blow molded vinyl. Imagine putting a tube-shaped balloon of vinyl into a mold and then inflating it. The balloon, when fully inflated, would

---

INTERESTING FACT

A common flaw with Playpal dolls is a compressed joint where the arm or leg meets the body. This defect is caused by the pressure that results from a doll being restrung too tightly.

take on the shape of the mold as the air forced it to expand outward. Due to this process, Playpal dolls are, thankfully, hollow and lighter weight as a result.

The Playpal dolls were first manufactured in 1959. Abraham Katz, Vice President of Ideal at that time, was in charge of doll design. Mr. Katz secured Neil Estern as a free-lance sculptor to create the Playpal family of dolls. Estern's wife, Anne, a costume designer, designed the outfits. Both strove for realism, trying to capture the essence of childhood in the 1960s. Mr. Estern's assistant was Vincent Di Fillipo, sculptor of the adorable Rub A Dub Dolly by Ideal and is still a popular doll artist today.

The 1959 spelling of Playpal was "Play Pal," changing to "Playpal" in 1960.

PATTI: Patti is the premiere member of the Playpal family, and the doll most people are familiar with. More Patti Playpal dolls were produced than other members of the Playpal family, making Suzy, Penny, Peter, and especially Johnny, Bonnie and Pattite harder to find.

Patti is 35 inches tall, the size of a 3 year old. She is marked "Ideal Toy Corp/ G 35" OR "B-19-1" (on her head) and "Ideal" in an oval on her back. It is believed that the early Pattis, manufactured in 1959, had what collectors refer to as "swivel" or "twist" wrists and were non-walkers. It is also believed that the Patti Playpal dolls that were walkers were manufactured in 1960 and beyond. There were reissues of the Patti Playpal dolls in 1981 and 1986.

INTERESTING FACT

Although it has been otherwise printed, Ideal was not the first toy company to produce a "life-size" doll. As you can see in the following photo, this 1953 American Character ad shows the Life Size Sweet Sue doll. Ideal was, however, a trail blazer in manufacturing dolls and toys that had celebrity and promotional tie-ins.

PETER: Peter, with his unique brand of freckled face charm, is the most expensive of the Playpal dolls in today's collector's market. He is 38 inches tall, the size of a 4 year old. A rare 36 inch size was made so that salesmen could fit these Peter dolls in the trunks of their cars. In addition to being hard to find, Peters are special in the fact that boy dolls, in general, are hard to find. Peter dolls are marked © Ideal Toy Corp / BE - 35-38 (on head) and Ideal Toy Corp / W-38 / Pat pend (on body.)

PENNY: Penny was made for only one year, 1959. At 32 inches, she is the size of a 2 year old and has a soft, rounded face that is endearing to anyone that sees her. She is marked "Ideal Doll / 32 - E - L" OR "B-32 Pat. Pend." (on her head) and "Ideal" in an oval on her back.

SUZY: An angelic baby doll that is the size of a 1 year old, 28 inches. In ads of the era, her name is spelled both Suzy and Suzie. She came in both straight and curly hair, with curly being the most common. Suzys are marked "Ideal Doll / O.E.B. - 28 - 55" OR "24 - 3" (on head) and "Ideal Toy Corp B 28" OR "Ideal" in an oval on the back.

TWINS, BONNIE and **JOHNNY**: These dolls are exceptionally hard to find, with Johnny being the harder of the two. Bonnie has rooted hair and Johnny is the only Playpal doll with painted hair. They are 24 inches long, making them the size of 3 month old babies. Johnny is marked "Ideal Doll / BB - 24- 3" (on head), "Ideal" (in oval on body). Bonnie is marked "Ideal Doll / OEB - 24 - 3" (on head) and Ideal (in circle) 23 (on body)

PATTITE: Pattite is simply a miniature version of the Patti Playpal doll. She is sought after by collectors and roundly prized. At 18 inches, she is a solid, substantial doll that is a joy to hold and handle! She is marked "Ideal Toy Corp / G - 18" on head.

THE 36 INCH SHIRLEY TEMPLE: This doll has the same body mold as Patti, but with a different head. Although she is not officially a "Playpal" there is no doubt that Playpal collectors scramble for her and that she brings top dollar at auction. If she has lost her original moppet curls, look for the "Ideal" marking on her back. In any shape, she is a valuable doll!

The Beauty of Companion Dolls

During telephone conversations with friends of mine who, like myself, are mothers of 3 year olds, I have often been asked why it is that the conversation is constantly being interrupted by their 3 year old, while my end of the line is always quiet and peaceful. The answer is simple, really. My 3 year old happens to be a set of twins. By virtue of the fact that there are two of them, they tend to keep each other occupied and amused and depend less on Mom. With that in mind, it is easy to see the value of a life-size companion doll that can give children a few more options than a baby doll can. Baby dolls can be diapered and fed only so much. But "someone your age" can do all the things you can - watch TV, play dress-up, or hold one end of the jump rope. This was the real beauty of the 60s companion dolls - 3 and 4 year olds could have a friend or "twin" and 5-9 year olds could have a little brother or sister!

A Rockette lineup of mint Patti Playpals. *Courtesy of Tara Wood.*

Joanie, Patti and Suzy, visiting Aunt Willa.
Dolls courtesy of Willa Moore.

A roundup of "real" dolls, with Taylor Gabel in front of her mom's Playpal collection.

My daughter, Tiffin in 1991 with her Patti Playpal who wears one of her kindergarten dresses.

Neil Estern, Sculptor

I don't consider Neil Estern a doll maker. In my opinion, he made children. For someone, who, in his own words, never grew up around dolls, sisters or even girl cousins, he managed to sculpt the very heart and essence of childhood in the form of a family of dolls. Patti Playpal is the image of everyone's ideal child. Not everyone's ideal doll, but their ideal child. Her face says it all. She is not only beautiful, with a fresh as paint brightness to her facial coloring and a rich brilliance in all of her many hair colors - she is also the very image of serenity (a trait that anyone would value in a child). Her tranquil beauty and her realism takes people's breath away. She is a 35 inch monument to childhood.

It was Abe Katz from the Ideal Toy Company who asked Mr. Estern to create a 36 inch doll that could wear real children's clothing. From that simple concept came the works of art that we now know as the Playpal family. Patti herself made quite the impression on the little girls of the early 60s, and was, overall, so successful that a Patti Playpal balloon was made for the Macy's Thanksgiving Day Parade.

Neil's wife, Anne Estern designed the early Playpal outfits and hairstyles, including the Red and White Checked Pinafore Dress, the Blue and White Checked Sailor Dress, the Green Apron Dress and the Organdy Pinafore Dresses. Some variations to the designs were later added by Mary Bauer, in the form of added lace, appliques and curlier hairstyles. Patti sported a very smart wardrobe that was representative of the best that the 1960s had to offer.

Other dolls to Mr. Estern's credit are Crissy, Kissy, Giggles, Betsy Wetsy, and Shirley Temple. Both Neil and Anne together thought up and designed the Flowerkin dolls and Puppetrina for the Goldberger Toy Company.

Also to his credit, Mr. Estern has sculpted the 8'11" seated statue of FDR and the 7'4" standing statue of Eleanor Roosevelt in the FDR Memorial - Washington, D.C., the 6'7" standing statue of Fiorello La Guardia for the La Guardia Memorial Park, New York, a portrait bust of JFK for the Kennedy Memorial, New York, and many other sculptures too numerous to mention. He has been the recipient of numerous awards and honors and is included in *Who's Who*, *Who's Who in American Art*, *Who's Who In The East*, and *Who's Who in The World*.

And as I said in my letter to him - Thank you, Mr. Estern for giving us these beautiful dolls. Sculptures in public places are certainly more of a challenge and will be gazed upon and admired by large numbers of people, but they will never be loved as family members, as these dolls were.

Chapter Two
PATTI PLAYPAL

Original outfits, hair colors and styles, 1981 reissue, 1986 reissue

PATTI PLAYPAL FACTS AT A GLANCE

Patti is about 35 inches tall (some Pattis measure in at 36 inches, probably due to hair style variations)
She was made to resemble a 3 year old child.

HAIR COLORS:

Blonde (most common)
Auburn
Brunette (also called dark brown or black)
Brown (a lighter brown than brunette - rare)
True Black (rare)
Black Cherry (standard brunette with a reddish cast - very rare)
Strawberry Blonde (standard Patti blonde with a red tint to it)
Light Auburn or **Cinnamon** (rare)
Carrot Top (rare)
Platinum (rare)
Gold Platinum or **Champagne** (rare)

EYE COLORS

Blue
Blue-Green
Green

> ## INTERESTING FACT
> Suzy Playpal with straight hair is considered rare.

HAIR STYLES:

Long straight hair with straight bangs (most common)
Curly Bangs (same hair style but with curly bangs - rare)
Flip (long straight hair, but with curly bangs and curly ends - rare)
Curly Top (tightly curled hair, usually about chin length)
Curly Bob (like the Curly Top, but very short - rare)
Ponytail (hair is rooted all around the hair line to accommodate the ponytail - very rare)
Pulled-back (rare)
Pageboy (like the long straight hairstyle, but a shorter version, about chin to shoulder length. Although these dolls are hard to find, they are not valued at more because this style would be very easy to "create" out of the more common long, straight style.)
No Bangs Patti (rare)
Spitcurl (rare)

Patti Playpal Hair Colors

Brown. Courtesy of Jill Cohen.

Blonde. Courtesy of Ray & Anne Geeck.

Auburn

Brunette (Also called dark brown or black)

True Black (rare)

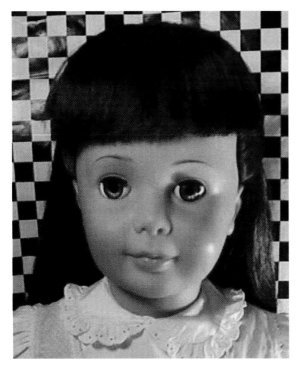

Black Cherry (Standard Brunette with a reddish cast, very rare) *Photography by John Meideiros.*

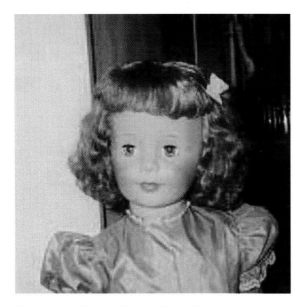

Strawberry Blonde (Standard Patti Blonde with a red tint to it)

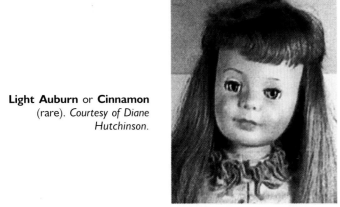

Light Auburn or **Cinnamon** (rare). *Courtesy of Diane Hutchinson.*

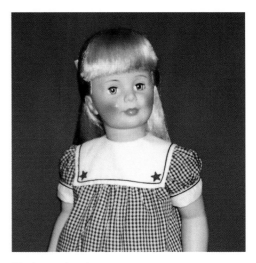

Carrot Top (rare)

Platinum (rare)

Gold platinum. *Courtesy of John Sonnier.*

Patti Playpal Hair Styles

Long straight hair with straight bangs (most common)

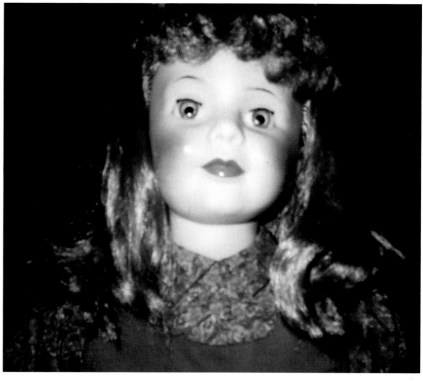

Flip (long straight hair, but with curly bangs and curly ends - rare)
Courtesy of Tara Wood.

Curly Bangs (same hair style but with curly bangs - rare)

Curly Top (tightly curled hair, usually about chin length)
Courtesy of Kathy Hosteller.

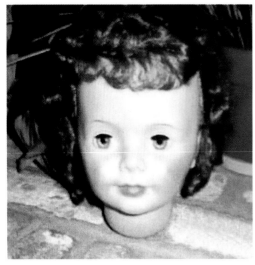

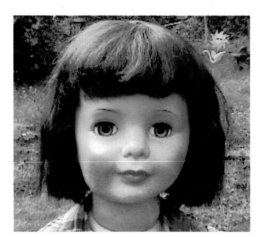

Pageboy (like the long straight hairstyle, but a shorter version, about chin to shoulder length. Although these dolls are hard to find, they are not valued at more because this style would be very easy to "create" out of the more common long, straight style.) *Courtesy of Miss Nancy.*

Curly Bob (like the Curly Top, but very short - rare) *Courtesy of Judie Conroy.*

Ponytail (hair is rooted all around the hair line to accommodate the ponytail - very rare)

Pulled-back (rare)

No Bangs Patti (rare) *Courtesy of Kathy Hosteller.*

Spitcurl (rare)

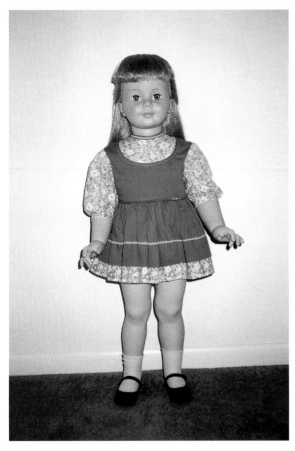 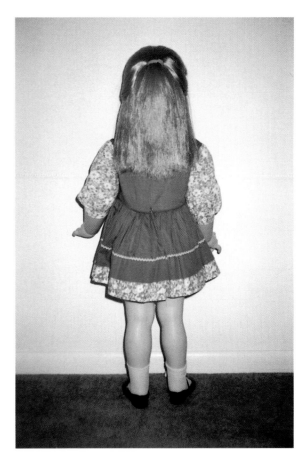

A great example of the "Pulled-Back" hair style as shown on this wonderful Carrot Top Patti. *Courtesy of Murray Hilliard.*
VALUE: IN MINT CONDITION, DOLL ONLY, NO ORIGINAL OUTFIT - CARROT TOP PULLED BACK HAIR STYLE: $850.

Example of "No Bangs" Patti, with side tendrils from the collection of Cindy Wilson, Lewiston, Idaho.
VALUE: IN MINT CONDITION, DOLL ONLY, NO ORIGINAL OUTFIT - NO BANGS PATTI: $350

Example of the "Curly Bangs, Straight Hair" hairstyle, in the rare cinnamon (or light auburn) color. *Courtesy of Zaundra Rickson.*
VALUE: IN MINT CONDITION, DOLL ONLY, NO ORIGINAL OUTFIT - CINNAMON CURLY BANGS: $400

Examples of "Pageboy" hairstyle found on Patti Playpals. The style is just like the standard Long Straight hairstyle, but shorter in length. No special value has been established for these dolls, because a Pageboy hairstyle can easily be created. *Doll from the estate of Rosalie Grigsby, represented by Esther Williams. Doll with bows courtesy of Zaundra Rickson*

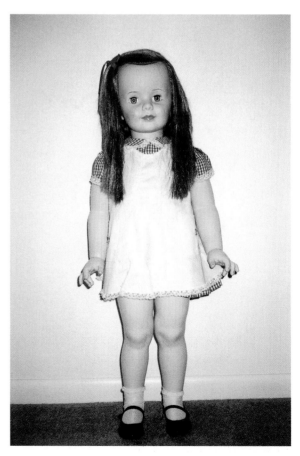

Two additional examples of the "No Bangs" Patti, a rare Playpal hairstyle. The auburn on the right is courtesy of Murray Hilliard and the Cinnamon on the left is courtesy of Diane Simmons. VALUE: IN MINT CONDITION, DOLL ONLY, NO ORIGINAL OUTFIT - AUBURN - 375 CINNAMON: $450

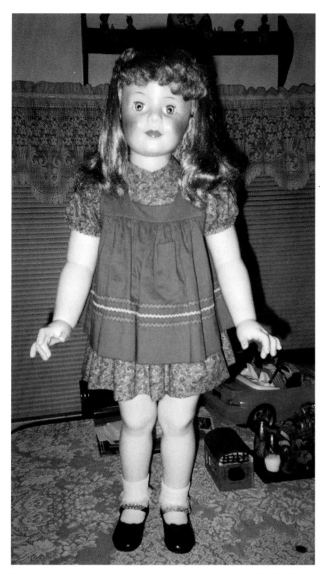

Example of the hairstyle collectors refer to as the "Flip." The Flip is the same as the "Curly Bangs" hair style, which consists of curly bangs with long straight hair, only the Flip style pairs the curly bangs with a curl on the ends of the hair - which usually curls up. *Courtesy of Tara Wood.*
VALUE: IN MINT CONDITION, DOLL ONLY, NO ORIGINAL OUTFIT - STANDARD BLOND FLIP: $300

A beautiful example of an Auburn Spitcurl. When redressed in vintage clothing, Playpal dolls can be just as striking as when originally dressed. *Courtesy of Pat Vaillancourt.*
VALUE: IF IN MINT CONDITION, DOLL ONLY, NO ORIGINAL OUTFIT: $375 IF IN FAIR CONDITION, ABOUT $225

INTERESTING FACT

In the first appearance that the Playpal dolls made in the 1960 Sears Christmas catalog, they were listed not as Playpals, but as "Honey Mates." Both Patti and Peter were featured in the catalog and the outfits that they wore, red crested blazer for Peter and peasant-style dress for Patti are referred to by collectors as the "Honey Mate" outfits, and are considered hard to find.

Close up detail of the "Pulled Back" and the "Ponytail" hairstyles. The Pulled Back style is rooted right at the hairline, with longer strands of hair on either side meant to be pulled back. The Ponytail dolls have a part in their hair, about half way back, with wispy bangs-like shorter hair that frames the face. *Courtesy of John Sonnier.*
VALUE: IN MINT CONDITION, DOLL ONLY, NO ORIGINAL OUTFIT - PLATINUM PATTI: $750, CARROT TOP: $750

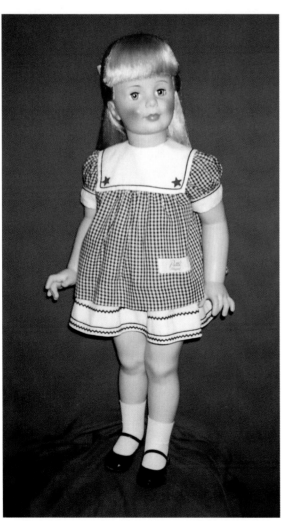

A wonderful example of the rare Platinum hair color (L). This, and the Carrot Top Pony Tail (R) are from the collection of John Sonnier. VALUE: IN MINT CONDITION, DOLL ONLY, NO ORIGINAL OUTFIT - PLATINUM: $750, CARROT TOP: $750

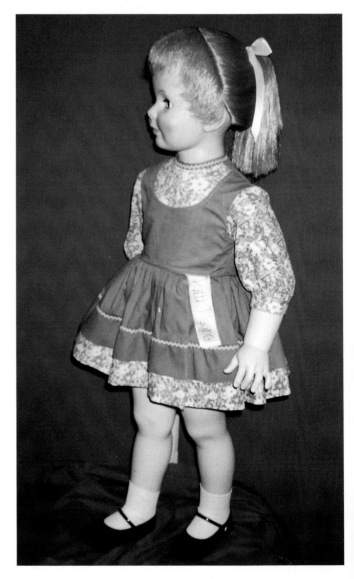

A 1960 Patti with a gorgeous example of the rare carrot top hair color. When mint, the sheen on these carrot tops is outstanding. *Courtesy of Tara Wood.*
VALUE: IN MINT CONDITION, DOLL ONLY, NO ORIGINAL OUTFIT - CARROT TOP: $750

This is a stunning example of the Gold Platinum or "Champagne" blonde Patti Playpal hair color. It is similar to the Platinum color, but with a distinct gold cast to it and notably different in color than a Standard Blonde Patti. This hair color is extremely hard to find. Another unusual thing about this Patti is that she has hazel eyes, like the ones used in the 36 inch Shirley Temples, and she has very pale blue eye shadow. *Courtesy of John Sonnier.*
VALUE: IN MINT CONDITION, DOLL ONLY, NO ORIGINAL OUTFIT - GOLD PLATINUM PATTI: $400

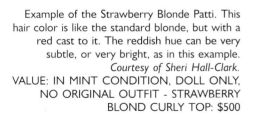

Example of the Strawberry Blonde Patti. This hair color is like the standard blonde, but with a red cast to it. The reddish hue can be very subtle, or very bright, as in this example. *Courtesy of Sheri Hall-Clark.*
VALUE: IN MINT CONDITION, DOLL ONLY, NO ORIGINAL OUTFIT - STRAWBERRY BLOND CURLY TOP: $500

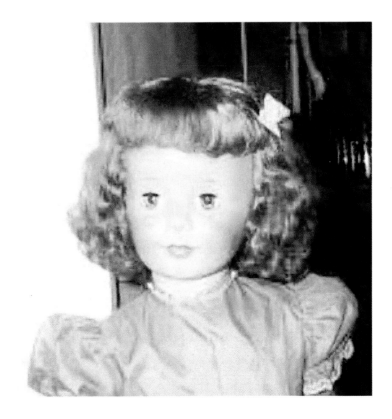

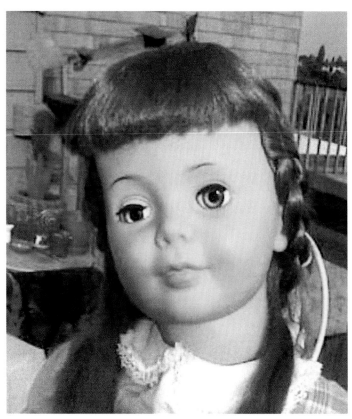

This is an example of the Light Auburn or "Cinnamon" Patti Playpal hair color. Some collectors will refer to this color as "strawberry blonde," but the Patti Playpal strawberry blonde is a standard blonde with a red blush to it. Sometimes this red cast can only be seen well in the sunlight, but it is a distinctly different hair color from the standard blonde, and very hard to find. *Courtesy of Suzanne Vlach.*
VALUE: IN MINT CONDITION, DOLL ONLY, NO ORIGINAL OUTFIT - CINNAMON PATTI: $300

Another Champagne or Gold Platinum Patti. This doll's hair could use a washing, and is actually more representative of the type of condition that you may find a gold platinum in. This photo may help you recognize this rarity if found in less than mint condition. It is reported that a doll hospital in Texas had 7 of these "Gold Platinum" Patti Playpal heads, obtained after the manufacture of the Playpal dolls, as extra parts. All 7 of these heads had pale blue eye shadow. *Courtesy of Karen Adams.*
VALUE: IN FAIR CONDITION, DOLL ONLY, NO ORIGINAL OUTFIT - GOLD PLATINUM PATTI: $400

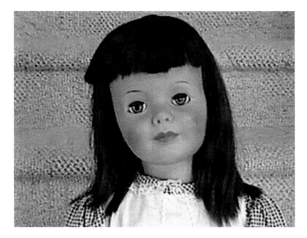

An example of a Patti with True Black hair. *Courtesy of Good Earth Collectibles - Middleburgh, NY.*
VALUE: IN MINT CONDITION, DOLL ONLY, NO ORIGINAL OUTFIT - TRUE BLACK HAIR PATTI: $550

Although it does not translate perfectly through printing, this is an example of the "Black Cherry" Patti Playpal hair color. Black Cherry is a standard brunette with a reddish cast to it. This hair color and texture is similar to that of the Lori Martin doll. Thank you, John Medeiros. *Photography by John Medeiros.*
VALUE: IN MINT CONDITION, DOLL ONLY, NO ORIGINAL OUTFIT - BLACK CHERRY PULLED BACK HAIRSTYLE: $350

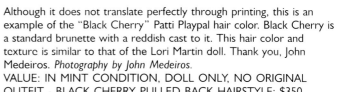

A Cinnamon Curly Top (LEFT) and Brunette Curly Top (ABOVE) - two real prizes! *Photography by John Medeiros.*
VALUE: IN MINT CONDITION, DOLL ONLY, NO ORIGINAL OUTFIT - CINNAMON CURLY TOP: $350 BRUNETTE CURLY TOP: $275

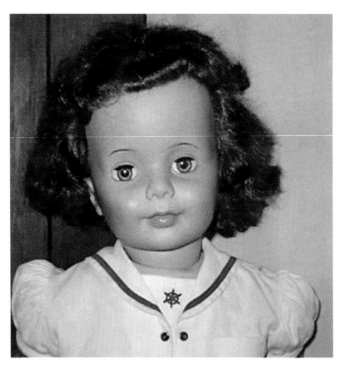

Example of Auburn Curly Top. The Curly Top is a longer cut than the very rare Curly Bob. *Courtesy of Ray and Anne Geeck.*
VALUE: IF IN MINT CONDITION, DOLL ONLY, NO ORIGINAL OUTFIT - AUBURN CURLY TOP PATTI: $275. IF IN FAIR CONDITION, ABOUT $175

Astonishing Auburn Pageboy. *Courtesy of Karen Adams of Scottsdale Antiques.*

Said to be a factory glitch, this beautiful mint Patti has one blue and one brown eye. *Courtesy of Judy Borges.*
NO VALUE HAS BEEN ESTABLISHED FOR THIS DOLL

A Playpal collector, Liz Opio, e-mailed this story to me:

As far as I can remember, the greatest pleasure I had in childhood was receiving a doll. Even today, as an adult, that special excitement and utter joy can return me to infancy. You are only as old as you feel and I feel wonderfully youthful!

My first encounter with Patti Playpal was on December 16, 1967, my fifth birthday. My Mom picked both my sister and me up from the babysitter after a long day at work. When we arrived home, Mom gave me a jar full of candy and a mysterious box way too large for me to open. With the help of my sister and Mom I was finally able to open the box, and out came Patti. Upon first looking at her, I was horrified. I had never seen a doll so big - she was almost my size!

For days, I would not go near Patti. Mom would try to coax me into playing with her by throwing her voice pretending to be Patti talking to me. Not a very good idea - even more frightened, I would run away screaming.

Over time, I mustered up the courage to actually walk over to Patti, only to stare at her, still too scared to touch. I was attracted to her beautiful long blonde hair. Eventually, I picked up a brush and began brushing it. At first, I was afraid that Patti would come to life and start yelling at me for brushing too hard, but that fear quickly disappeared. We slowly became

the best of friends and I would take her everywhere my Mom would allow me to. We took baths together and I was thrilled that I could lend her my clothing! In moments of tantrums, Patti received lots of abuse. In one of my beautician experiments, I cut her hair and painted her face. Today, I am a hair designer and makeup artist be trade, and have been since 1982. Finally, after years of tough love, Mom got rid of Patti.

A couple of years ago, I read an article on the Playpal family in a doll magazine. I was again yearning to have and hold a Patti. Only this time I would impart her much deserved loving care. I bought my first Patti just weeks after reading the article.

A year later, I located a brunette in her original dress and shoes. Of all the dolls that I have in my collection, ranging from antique, cloth, porcelain artist edition and vintage dolls, Patti will always have a very special place in my heart!

Dawn Nielsen Dinegan from Oxford, Wisconsin with her prized Patti, Christmas, 1962.

BASIC VALUES FOR ORIGINAL PATTI PLAYPAL DOLLS

The following values are for dolls in MINT condition, DOLL ONLY - no original outfit, and for dolls with the standard "Long Straight" hairstyle, with the exception of the Carrot Top.

Standard Blond	$200
Auburn	$225
Brunette	$225
Brown (see description)	$350
True Black	$550
Black Cherry	$250
Strawberry Blond (see description)	$450
Light Auburn or Cinnamon	$300
Carrot Top	$750
Platinum	$750
Gold Platinum	$400

Patti's hairstyle can add quite a bit to her value. Please familiarize yourself with these dolls and become adept at determining the difference between a tampered with and a factory set.

Curly Bangs	add $100
Flip	add $100
Pulled back	add $100
	except for Platinum and Gold Platinum, for which this is a standard hairstyle
Curly Top	add $50
Curly Bob	add $100
Ponytail	add $150,
	except for Carrot Tops, for which this is a standard hairstyle
Pageboy	adds no value
No Bangs	add $150
Spitcurl	add $150

The very first Pattis, with the twisted wrists came in this box. Later dolls came in boxes that specified the dolls were "walkers." *Courtesy of John Sonnier.*

Brunette haired Curly Top. *Courtesy of Lori Anne Gabel.*
VALUE: IN MINT CONDITION, DOLL ONLY, NO ORIGINAL OUTFIT - BRUNETTE CURLY TOP: $275

Example of jointed or "swivel" wrists on the early Patti Playpals and the 36 inch Shirley Temples.

INTERESTING FACT

Most collectors concur that the 1959 Pattis were non walkers with swivel wrists, and that the 1960 Pattis were walkers with stationary wrists.

Inside view of strung Playpal arm.

An example of a green-eyed Patti. *Courtesy of Janet Porkrinchak.* VALUE: IN MINT CONDITION, DOLL ONLY, NO ORIGINAL OUTFIT - AUBURN PATTI: $225

This is an extremely rare prototype Patti Playpal from the collection of Murray Hilliard. This doll has "Follow me" eyes, similar to those on American Character's Toodles. She is all original in a ricrac pinafore with a pinwheel underdress. A value has not been established for this doll.

CANADIAN PATTIS

Just like Mattel's Chatty Cathy, Patti Playpal had Canadian counterparts made during the same time frame that she was popular in America. These Canadian "sisters" in the case of both the Chatty Cathy dolls and the Patti Playpals have some interesting variations in things such as hair color and face paint that are fascinating to the advanced collector! The Canadian Patti Playpals are harder to come by than the Canadian Chattys. They are marked with the name "Reliable" molded into their backs. Reliable Toy Company also made Ideal dolls for the Canadian market such as Tammy, Kissy and Misty. Canadian Pattis have subtler face paint than the American Pattis, and all seem to have the pinwheel-type eyes in blue and a soft coral lip color. The vinyl used in the Canadian Patti's heads is softer and more pliable. Their original outfits seem very similar to the American Playpal outfits. For the seasoned Playpal collector, obtaining these Canadian Pattis is an interesting challenge that can turn up more than a few surprises.

VALUE - VALUES HAVE NOT BEEN ESTABLISHED FOR CANADIAN PLAYPALS.

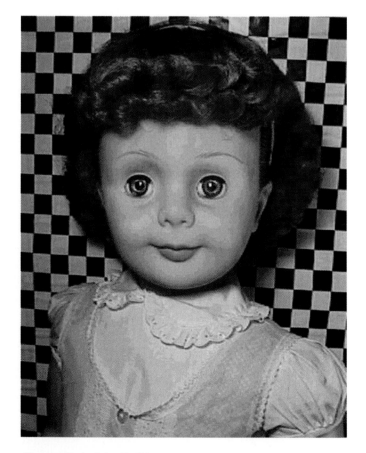

Photography by John Madeiros

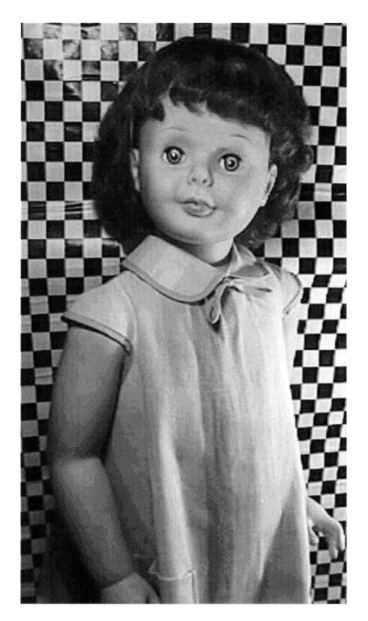
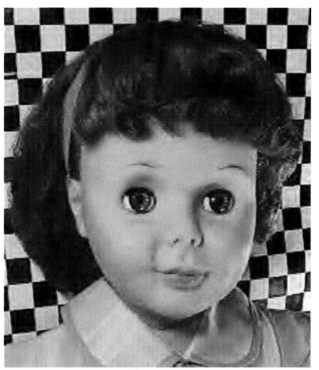

This Canadian Patti, with her unusual bright red lips, illustrates just the type of variation that makes the Canadian dolls so much fun. She is wearing an original outfit. *Photography by John Medeiros.*
VALUE - VALUES HAVE NOT BEEN ESTABLISHED FOR CANADIAN PLAYPALS

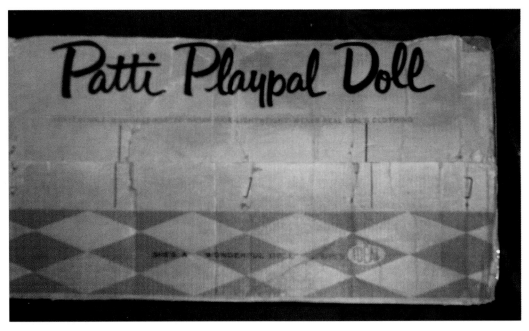

The earliest Pattis came in this box. Later Pattis came in boxes that specified that the dolls were "walkers." *Courtesy of John Sonnier.*

Later Patti Playpal box, probably from about 1960 on, with the words "She Walks." *Courtesy of Rose Korobanov.*

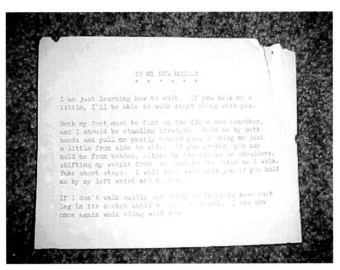

This is the note to "new mothers" that came with the Patti Playpal walkers. "I am just learning how to walk. If you help me a little, I'll be able to walk right along with you..." *Courtesy of Rose Korobanov.*

Patti Playpal Original Outfits

Doll clothing is often a record of the styles and preferences of a certain era. Patti's smocks, pinafores, jumpers and Mary Jane party shoes really showcase the late 1950s - early 1960s fashion.

Throughout the Playpal family, the use of fabrics such as dotted Swiss and organdy also captures a bygone era which valued both quality and femininity.

"Generic" Outfits for Playpals

During the time that the Playpal dolls were popular, various companies made clothing for the life sized dolls. Some of these outfits were available through the Christmas toy catalogs, and some were sold in toy stores.

Although this clothing was not made by Ideal, some Playpal collectors like to document these different outfits and include them in their collections.

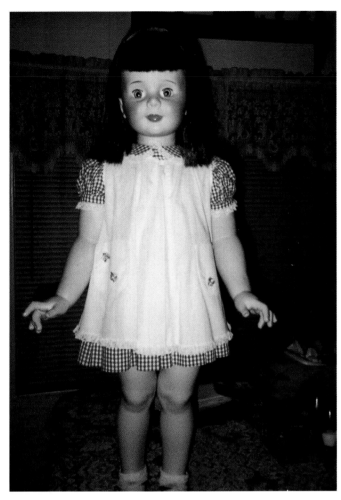

THE RED CHECKED - WHITE PINAFORE DRESS This 1959 brunette Patti is wearing one of the more frequently found original Playpal dresses, which consists of a red checked waisted cotton dress with a white cotton A - line pinafore. *Courtesy of Tara Wood.* VALUE - DRESS ONLY - WITHOUT UNDERWEAR OR SOCKS, IN EXCELLENT CONDITION - ABOUT $75

INTERESTING FACT

"Generic" outfits sold in the Sears catalog for Playpal sized dolls for between $3 and $4.

Comparison of 1982 Reissue Patti Playpal (left) and the original (right) Red Check/ White Pinafore Dress. The 1982 dress is all in one piece with an inset or "mock" pinafore. Red, white and green strawberries are stamped on the right and left panels of the pinafore. The original dress, from the late 50s - early 60s time frame is a red and white check cotton dress with a natural waistline and a detachable white cotton pinafore. The pinafore is found with and without flower appliques on the pockets and a rare version has red "Patti Playpal" stitching on the pocket. The red and white check underdress is found with and without banded or "cuffed" sleeves and with and without a black ribbon bow at the neckline. *Author's collection.*

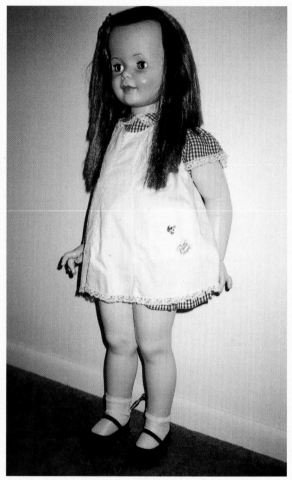

Two variations of this same dress shows a non-banded sleeve and a very hard to find example with red pocket stitching, which says "Patti Playpal," as modeled by this auburn No Bangs Patti. *Left, courtesy of author's collection; right, courtesy of Murray Hilliard.*
VALUE - DRESS ONLY - WITHOUT UNDERWEAR OR SOCKS, IN EXCELLENT CONDITION - ABOUT $75

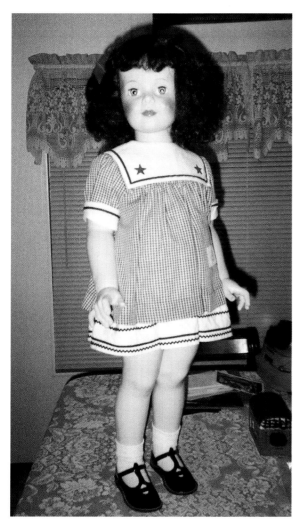

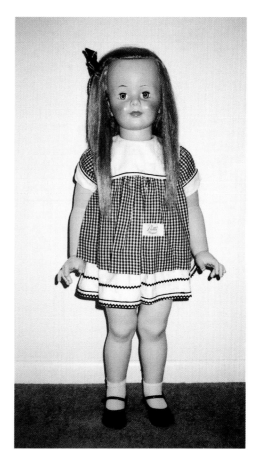

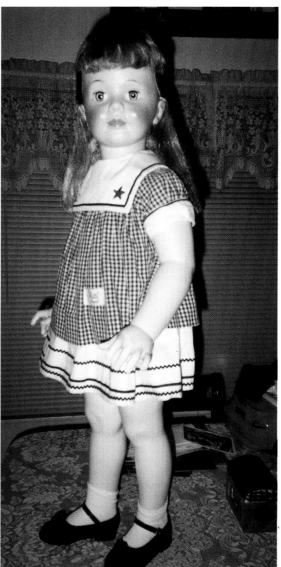

THE BLUE CHECKED SAILOR DRESS (ABOVE AND TOP RIGHT)
This 1959 black haired Curly Top Patti (left) is wearing the original blue checked Sailor Dress. *Courtesy of Tara Wood.* The No Bangs Patti on the right, also a 1959, is wearing a variation of the same dress, with a darker blue check. *Courtesy Murray Hilliard.*
VALUE - DRESS ONLY - WITHOUT UNDERWEAR OR SOCKS, IN EXCELLENT CONDITION - ABOUT $100

Right: A third variation of this dress shows a medium blue, but larger check overdress, that is shorter in length. *Courtesy of Tara Wood.*
VALUE - DRESS ONLY - WITHOUT UNDERWEAR OR SOCKS, IN EXCELLENT CONDITION - ABOUT $100

INTERESTING FACT

The quickest way to tell a 1981 reissue Patti Playpal from the original, at a glance, is that the 1981 versions have stationary, rather than sleep eyes.

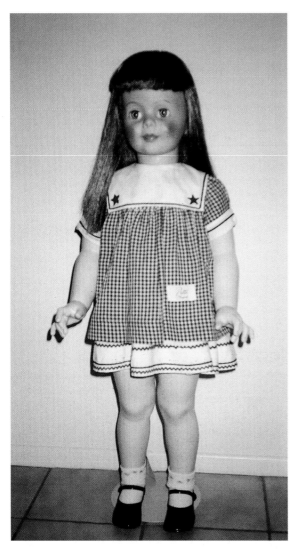
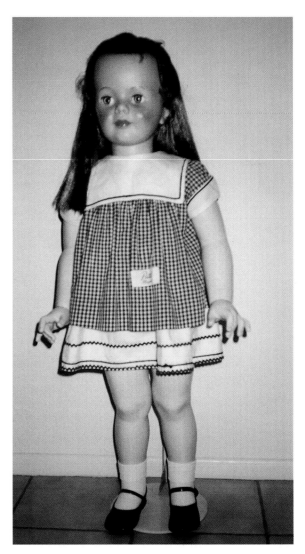

A look at these two Blue Checked Sailor Dresses shows variations in the collar appliques, the print color of the dress tag and in the length of the overdress. These are the type of variations that collectors love to notice! Small details of this type help to keep collecting interesting. *Both courtesy of Laura Brink.*
VALUE - DRESS ONLY - WITHOUT UNDERWEAR OR SOCKS, IN EXCELLENT CONDITION - ABOUT $100

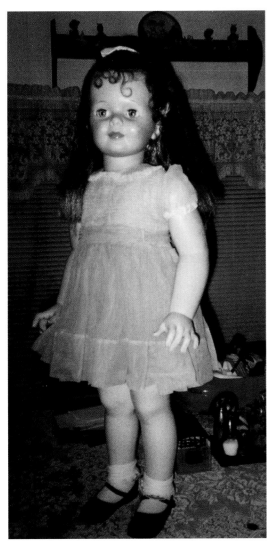

GREEN ORGANDY PARTY DRESS
A rare Spitcurl Patti with a very sweet face shows us the very rare and sought after Green Organdy Party Dress. The dress has very delicate pin tucks on the bodice and a lace-trimmed collar and cuffs. A slightly easier to find version of this dress comes in blue. Both the blue and the green versions of this dress, however, are considered rare. Thank you Tara Wood.
VALUE - DRESS ONLY - WITHOUT UNDERWEAR OR SOCKS - EXCELLENT CONDITION - ABOUT $200

THE GREEN APRON DRESS
This beautiful Carrot Top Patti is wearing the Green Apron Dress, which is a one piece green, white and orange print cotton dress with a dark green apron-look inset. *Courtesy of Lori Gabel.*
VALUE - DRESS ONLY - WITHOUT UNDERWEAR OR SOCKS, IN EXCELLENT CONDITION - ABOUT $100

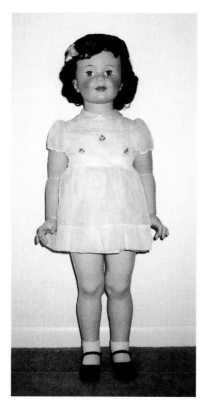

POWDER BLUE PINSTRIPE
A powder blue and white pin-striped dress is topped by a sheer organdy overdress with 3 flower appliques on the bodice. This dress, one of my personal favorites, is modeled by an absolutely striking black - haired Curly Top Patti. *Courtesy of Murray Hilliard.*
VALUE - DRESS ONLY - WITHOUT UNDERWEAR OR SOCKS, IN EXCELLENT CONDITION - ABOUT $125

Above: A slight variation of the Powder Blue Pinstripe Dress shows it cut fuller and longer. *Courtesy of Tara Wood.*
VALUE - DRESS ONLY - WITHOUT UNDERWEAR OR SOCKS, IN EXCELLENT CONDITION - ABOUT $125

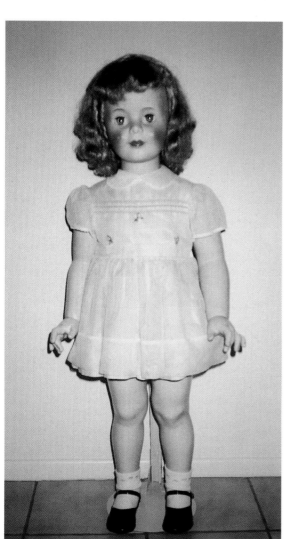

Left: Another slight variation of Powder Blue Pinstripe, with a blue/pink/blue ricrac pattern on bodice. *Courtesy of Laura Brink.*
VALUE - DRESS ONLY - WITHOUT UNDERWEAR OR SOCKS, IN EXCELLENT CONDITION - ABOUT $150

Organdy Pinafore Dresses

Several solid color as well as cotton print dresses of Patti Playpal were topped with the same white organdy pinafore. These dresses really showcase the era's classic styling for little girl's party wear.

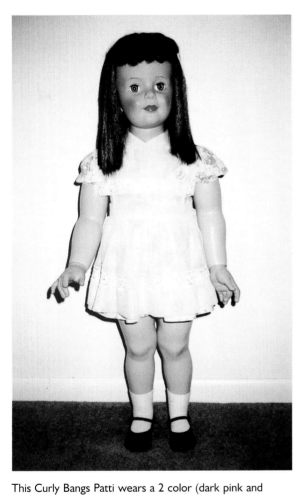

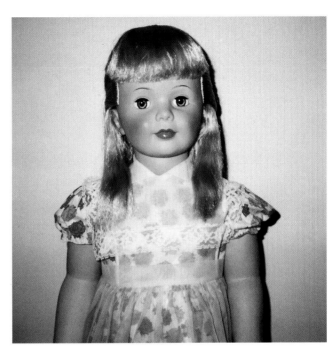

This platinum blonde Patti in the "pulled back" hair style is wearing the pink cabbage rose cotton dress with the sheer organdy pinafore. Courtesy *of Murray Hilliard.*
VALUE: IN MINT CONDITION, DOLL ONLY, NO ORIGINAL OUTFIT - PLATINUM PATTI: $750
VALUE - DRESS ONLY - WITHOUT UNDERWEAR OR SOCKS, IN EXCELLENT CONDITION - ABOUT $150

This Curly Bangs Patti wears a 2 color (dark pink and white) floral print dress under the organdy pinafore. *Courtesy of Murray Hilliard.*
VALUE - DRESS ONLY - WITHOUT UNDERWEAR OR SOCKS, IN EXCELLENT CONDITION - ABOUT $150

Two breathtaking Pattis in Organdy Pinafore Dresses, one with a solid color lilac underdress, and one with a lilac print pinafore. *Both courtesy of Lori Gabel.*
VALUE - EACH DRESS ONLY - WITHOUT UNDERWEAR OR SOCKS, IN EXCELLENT CONDITION - ABOUT $150

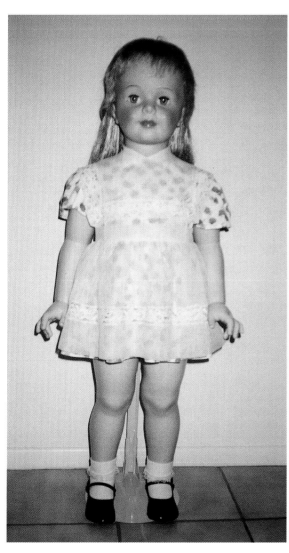
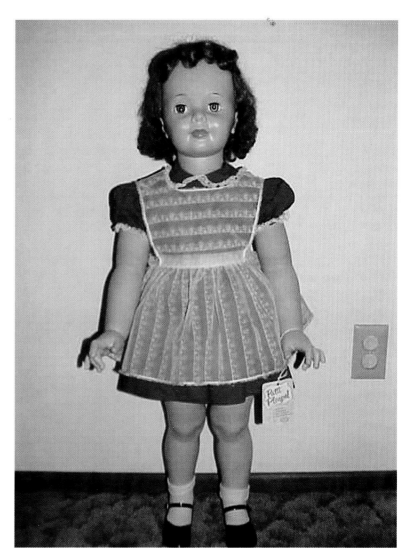

This rare Carrot Top wears an organdy pinafore set with a white background/lilac floral print underdress. *Courtesy of Laura Brink.*
VALUE - DRESS ONLY - WITHOUT UNDERWEAR OR SOCKS, IN EXCELLENT CONDITION - ABOUT $150

Beautiful and rare outfit which consists of a dark purple under dress and an unusual organdy pinafore. *Courtesy of Judy Borges.*
VALUE - DRESS ONLY - WITHOUT UNDERWEAR OR SOCKS, THIS ONE IN MINT CONDITION - ABOUT $200

Ricrac Pinafore Dresses

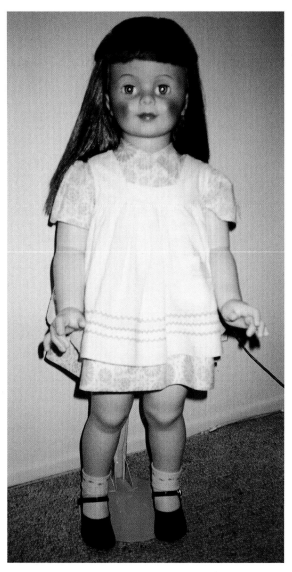

Various floral print dresses came paired with a white cotton pinafore, that had color coordinated ricrac at the hem. This is a light, pastel version of what collectors call the "pinwheel" floral print. The flowers in this print are perfectly round and resemble pinwheels. This pinwheel print came in at least 2 versions - hot pink/lime green and orange/yellow. *Courtesy of Laura Brink.*
VALUE - DRESS ONLY - WITHOUT UNDERWEAR OR SOCKS, IN EXCELLENT CONDITION - ABOUT $125

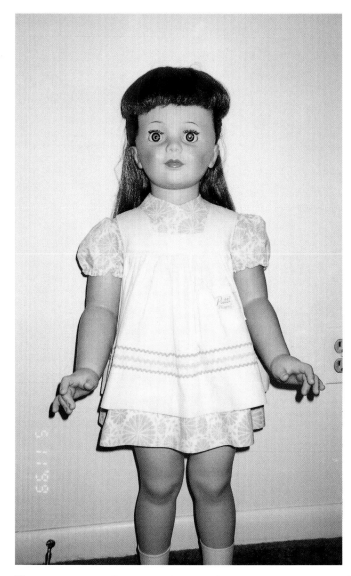

This rare prototype Patti wears the same dress, but with a variation of the ricrac - yellow/pink/yellow, instead of pink/yellow/pink. Thanks to the collection of M.H. Hilliard.
VALUE - DRESS ONLY - WITHOUT UNDERWEAR OR SOCKS, IN EXCELLENT CONDITION - ABOUT $125

Variation of the same dress with a very short pinafore. *Courtesy of Arlene Williams.*
VALUE - DRESS ONLY - WITHOUT UNDERWEAR OR SOCKS, IN EXCELLENT CONDITION - ABOUT $125
Please note - some advanced collectors may be willing to spend more for an unusual variation of this kind.

Another example of an underdress print variation on the ricrac pinafore outfit. *Photography by John Medeiros.*
VALUE - DRESS ONLY - WITHOUT UNDERWEAR OR SOCKS, IN EXCELLENT CONDITION - ABOUT $125

Velvet Jumpers

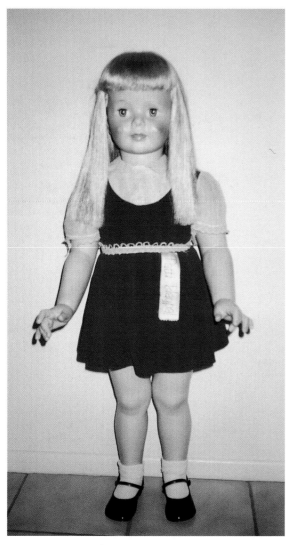

Above: This velvet jumper, in red, still has its original tag sewn into the waist of the dress and is shown on a rare platinum blonde Patti. These Velvet Jumpers, made for Patti Playpal, came in red, blue and green. *Courtesy of Laura Brink.*
VALUE - DRESS ONLY - WITHOUT UNDERWEAR OR SOCKS, IN EXCELLENT CONDITION - ABOUT $200

Top Right: This is the blue velvet version of the same jumper. *Photography by John Medeiros.*
VALUE - DRESS ONLY - WITHOUT UNDERWEAR OR SOCKS, IN EXCELLENT CONDITION - ABOUT $200

Right: BLACK VELVET CHRISTMAS DRESS.
This 1960 platinum blonde Patti shows off a dress as rare as she is, the Black Velvet Christmas Dress with white lace accents. *Courtesy of Tara Wood.*
VALUE - DRESS ONLY - WITHOUT UNDERWEAR OR SOCKS, IN EXCELLENT CONDITION - ABOUT $175

Plaid School Dresses

A double dose of rarity from collector Jill Cohen. A hard to find Brown Haired Patti in a super rare original outfit. This tagged dress, in polished blue cotton, is trimmed with tiny Spanish guitars and conga drums! For lack of original Ideal Company titles, I have named this little number the Ricky Ricardo Dress!
NO VALUE HAS BEEN ESTABLISHED FOR THIS DRESS

Plaid School Dress. This one piece cotton plaid school dress has been found on Patti Playpals. This particular dress was photographed on a Patti Playpal with the original owner on a Christmas day in the early 1960s. There is no attached slip and it is untagged. This dress could have been one of the "generics" available for Playpal sized dolls from Christmas catalogs and toy stores.
NO VALUE HAS BEEN ESTABLISHED FOR THIS DRESS

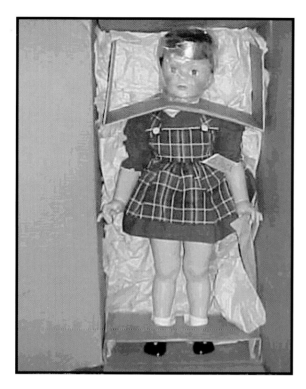

This Patti dress is a reverse of the usual Plaid School Dress, with a solid red underdress, and a plaid pinafore! And what an additional privilege to see a NRFB Patti!
Courtesy of Rose Korobanov.
VALUE - MINT, ABOUT $150

PLAID SCHOOL DRESS WITH RED APRON
This plaid school dress has a larger scale plaid which is red, white and blue, and a red apron with large navy buttons. This is not an apron inset, but an actual apron. *Courtesy of Lori Gabel.*
VALUE - DRESS ONLY - WITHOUT UNDERWEAR OR SOCKS, IN EXCELLENT CONDITION - ABOUT $125

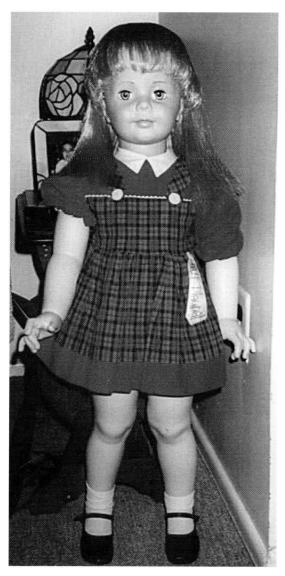

This is the same dress as the reverse plaid School Dress, but with a different color plaid apron. *Courtesy of Lori Gabel.*
VALUE - DRESS ONLY - WITHOUT UNDERWEAR OR SOCKS, IN EXCELLENT CONDITION - ABOUT $125

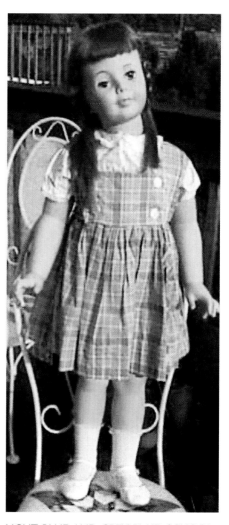

LIGHT BLUE AND GREY PLAID SCHOOL DRESS
Another of the plaid school-type dresses found on Patti Playpals and documented as original by original owners, this one in a light blue and grey plaid with a white cotton blouse. Similar fabric, construction and large round buttons as the other school dresses. *Courtesy of Suzanne Vlach.*
VALUE - DRESS ONLY - WITHOUT UNDERWEAR OR SOCKS, IN EXCELLENT CONDITION - ABOUT $125

INTERESTING FACT

PLAYPAL CLOTHING SIZES - Peter wears about a 4T, Patti a 3T, Penny a 2T, Suzy between a 12 and 18 month size, and the twins a 3 to 6 month size.

PINK ORGANDY DRESSES

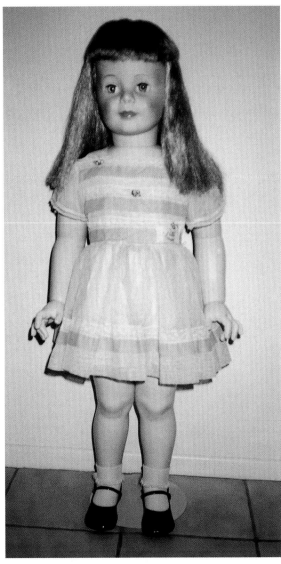

The Pink Organdy Dresses made for Patti Playpal were similar in style to the organdy dresses made for the 36 inch Shirley Temple. This dress, with its powder blue trim, is particularly hard to find. *Courtesy of Laura Brink.*
VALUE - DRESS ONLY - WITHOUT UNDERWEAR OR SOCKS, IN EXCELLENT CONDITION - ABOUT $175

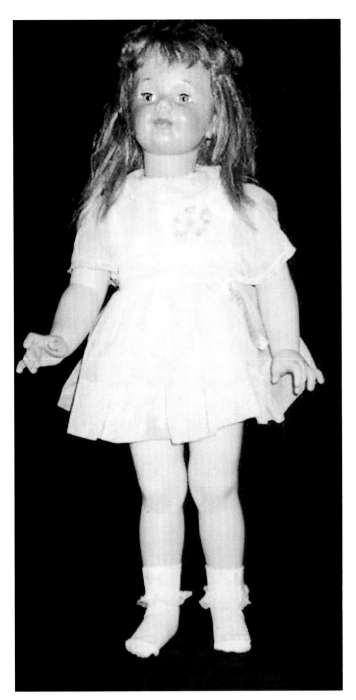

PINK ORGANDY
Original Patti Playpal dress, light pink organdy with flower applique on bodice. This was Marylou Stewart's childhood doll. This dress also comes in a variation that has a white bodice and a powder blue skirt. *Courtesy of Marylou Stewart.*
VALUE - DRESS ONLY - WITHOUT UNDERWEAR OR SOCKS, IN EXCELLENT CONDITION - ABOUT $150

This powder blue organdy dress is tagged, and appears to be an overdress. *Courtesy of Pat Vaillancourt.*

A most unusual and rare original Patti outfit, this dress is a "twin" to the Saucy Walker dress! Because of its rarity, the VALUE of this dress, MINT IS ABOUT $200. *Courtesy of JT9943.*

Sally Starr Outfit
This outfit was not made by Ideal, but was made especially for the Playpal dolls. Sally Starr was a 1960s TV cowgirl. Hat and boots were not originally included. *Courtesy of Janet Porkrinchak.*
VALUE - ORIGINAL AND EXCELLENT: $50

This is an example of the attached slip found on so many Playpal fashions.

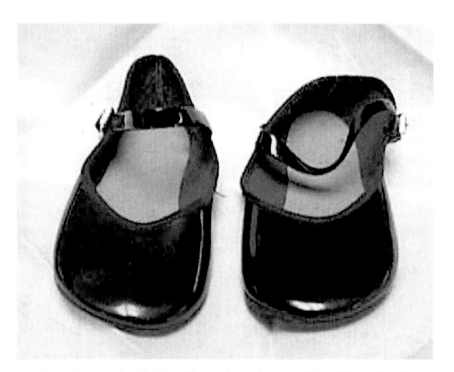

Great close up of original Patti Playpal shoes. *Courtesy of Velma Wilson Glass Hut.*
VALUE - IN EXCELLENT CONDITION - ABOUT $50

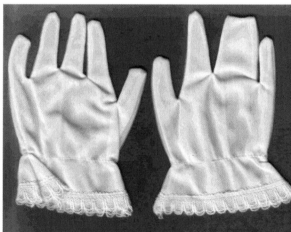

A very unique item - gloves made for Playpal sized dolls with a special adaptation for the 2nd and 3rd fingers. *Courtesy of Shirley Merrill.*
VALUE - about $20

1981 REISSUE PATTI PLAYPAL

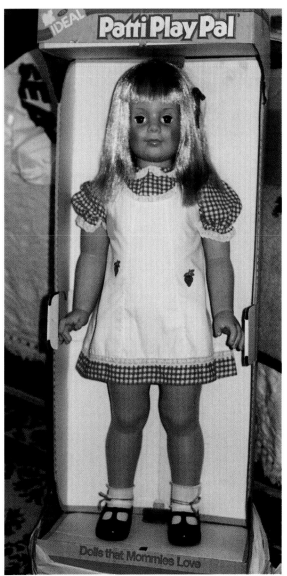

VALUES for 1981 Reissue
MIB White doll, about $125
MIB African American doll, about $175
1981 Patti Playpal - Mint in Box. Thank you to Pamela Payne and Caroline Steward

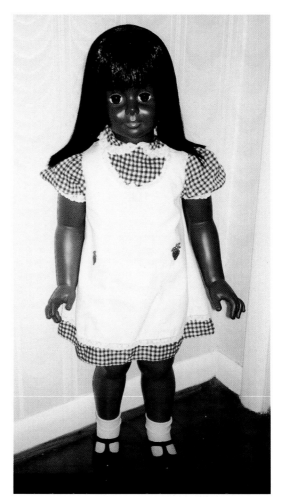

Above: Black version of the 1980s Patti Playpal *Courtesy of Valaina Maher.*

Top Right: This beautifully mint reissue Patti Playpal shows a variation of her original outfit, which is in a darker red check. *Courtesy of Debbie Garrett.*

Right: Slight variation of the 1981 African American Patti, with lighter skin and eye color. *Photography by John Medeiros.*

1986 Talking Patty Play Pal

In the 1980s large talking dolls like Crickett were becoming quite popular with consumers. It was during this time that Ideal created the Talking Patty Play Pal. Yes, Patty was spelled with a "y" and Play Pal went back to being 2 separate words, the way it was spelled in 1959, the very first year of production.

Talking Patty was a platinum blond, with vinyl arms, legs and head, and a stuffed cloth body. She spoke with the aid of a pink lunch box, which was actually a tape recorder. Separate sets of books and audio cassette tapes were sold so that Patty could tell a variety of different stories. The books also contain activities such as mazes, puzzles, paper dolls and tongue twisters for children to participate in, and the cassettes were synchromation tapes that caused Patty's mouth to move in conjunction with the words. Separate boxed outfits were available for Talking Patty Play Pal as well.

Although this Patty was not a companion-sized doll, many advanced collectors want her in their collection, because she bears the name, Patty Play Pal.

I have uncovered the following Patty Play Pal book and cassette tape sets for the 1986 Talking Patty Play Pal:

1) Adventure In Outer Space
2) A Play Pal Workout
3) A Voyage Into The Past
4) A Funtime Party
5) The Play Pal Detective Agency
6) A Fairytale Fantasy
7) The Play Pal Club
8) A Play Pal Slumber Party
9) Let's Go To The Beach

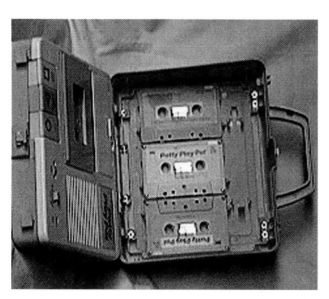

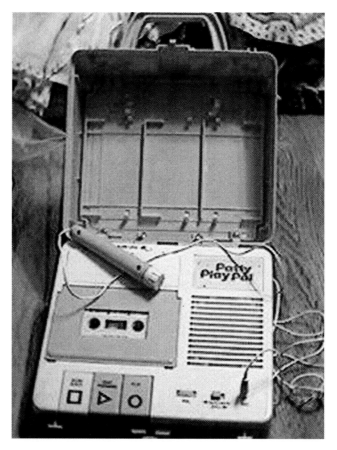

Values

1986 Talking Patty Play Pal - (without lunchbox): $50-$75.

1986 Talking Patty Play Pal Tape and Book Sets:
  Book and Tape Set - Mint and complete: $12- $20
  Individual books and Tapes: $5-$7 each

Outfits - Loose and incomplete - very good condition: $10-$20
Outfits - Mint in package: $25-$35
Lunchbox/ tape recorder - mint and complete: $50

Another of the books published in 1987 for the Talking Patty Play Pal, "The Patty Play Pal Detective Agency." Value: $5: $7.

1987 Patty Play Pal hardcover book and tape set, made by Ideal and sold to be used in conjunction with the Talking Patty Play Pal doll. This book contains craft projects, songs and "keep busy" activities for Patty and her young mommy. 24 pages, it is valued, mint in package at about $12: $20.

1987 TALKING PATTY PLAY PAL IN ORIGINAL PRETTY PARTY DRESS

I have uncovered the following original packaged outfits for the 1987 Talking Patty Play Pal doll:

Red and White Cheerleader Outfit
Pink and Purple Workout Suit
Powder Blue Princess Dress
Red Satin Party Dress
Lilac Pajamas

One of the 5 original outfits available for the Talking Patty Play Pal doll, #8715 The Pretty Party Dress. Satiny red fabric dress with white sash and bow at waist. Came with matching hair bow and cassette.

Original tagged outfit for the 1987 Talking Patty Play Pal.
Courtesy of Jill Simpson.
VALUE - OUTFIT ONLY, FAIR CONDITION: $10

Original Talking Patti Playpal sweatshirt. *Courtesy of Maria Love.*
VALUE - EXCELLENT: $10

CHAPTER 3.
PLAYPAL FAMILY MEMBERS
(Pattite, Peter, Penny, Suzy, Twins Bonnie and Johnny)

PATTITE

A veritable gold mine of Pattites. Imagine the amount of effort and searching that must go into locating and obtaining so many of these extremely rare dolls!! *Courtesy of the M. Hilliard Collection.*
VALUE - PATTITE IN MINT CONDITION AND ORIGINAL OUTFIT: $600: $700
PATTITE IN MINT CONDITION WITHOUT ORIGINAL OUTFIT - DEDUCT ABOUT $150

PATTITE - FACTS AT A GLANCE

The beautiful Pattite doll is considered a miniature version of the Patti Playpal. She was made in 1960, and again in 1961, as a walker version.

At 18 inches, she is a solid, substantial doll that is a joy to hold and handle! She is marked "Ideal Toy Corp / G - 18" on head.

Hair colors: Blond, Brunette, Auburn

Some of Pattite's Original Outfits:

1) Dark pink cotton dress with white dotted Swiss apron
2) Light green cotton dress with matching jacket (pictured)
3) Red and white checked dress with white pinafore (a miniature version of Patti's - pictured)
4) Mini Dot Dress with Over Blouse in either tan or gray. (pictured)
5) Pink Gingham (pictured)
6) Green Gingham / White Pinafore (pictured)

PLEASE - familiarize yourself with original Pattite dresses and with Ideal workmanship, as many Pattite clothing is not tagged, and very pricey.

VALUES FOR PATTITE

VALUE - PATTITE IN MINT CONDITION AND ORIGINAL OUTFIT: $600: $700
PATTITE IN MINT CONDITION WITHOUT ORIGINAL OUTFIT - DEDUCT ABOUT $150

Three Pattites - a Blonde, a Brunette, and a Redhead, all in original outfits. *Courtesy of Tara Wood.*
VALUE - PATTITE IN MINT CONDITION AND ORIGINAL OUTFIT: $600: $700
PATTITE IN MINT CONDITION WITHOUT ORIGINAL OUTFIT - DEDUCT ABOUT $150

Original Pattite Outfit - Pink Gingham Dress.
Original Pattite Outfit - Green Gingham / White Pinafore. *Courtesy of Tara Wood.*
VALUE - ORIGINAL PATTITE DRESSES ARE WORTH ABOUT $150.

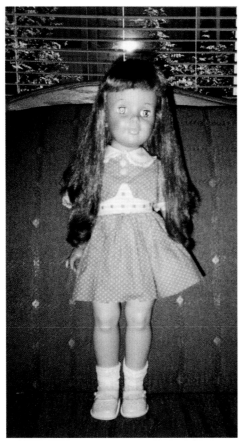

Original Pattite Outfit in tan - Mini Dot with Over Blouse. *Courtesy of Tara Wood.* Variation of same dress, in gray, rather than tan. *Courtesy of Kathy Ebey.*
VALUE - ORIGINAL PATTITE DRESSES ARE WORTH ABOUT $150.

This original catalog ad shows Pattite, center, in an original mint green dress.
VALUE - PATTITE IN MINT CONDITION AND ORIGINAL OUTFIT: $600: $700
PATTITE IN MINT CONDITION WITHOUT ORIGINAL OUTFIT - DEDUCT ABOUT $150

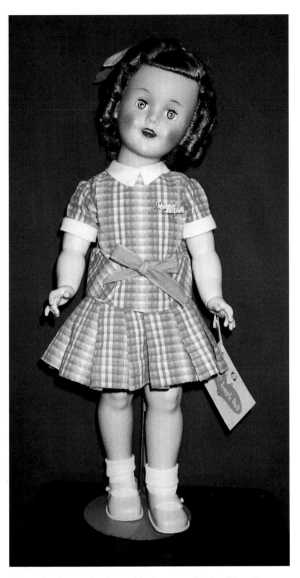

Using the Pattite body mold, this rare Shirley Temple Walker was made. *Courtesy of John Sonnier.*
VALUE ABOUT $1000

Lori Anne Gabel's prized Pattite, in original outfit, a miniature version of Patti Playpal's most common dress.

PETER

PETER PLAYPAL - FACTS AT A GLANCE

Peter, and his unique brand of freckled-face charm, is the most expensive of the Playpal dolls in today's collector's market. He was made in 1960 and 1961. He is 38 inches tall, the size of a 4 year old. A rare 36 inch size was also made. The rare 36 inch Peter Playpal "Salesman" doll came about as a result of Playpal traveling salesmen complaining that the original 38 inch versions were too tall to fit in the trunks of their cars. Thus, as it is told, Ideal shrunk these salesman demos down by 2 inches. These shorter Peters are too rare to even come up with a value on!

In addition to being hard to find, Peters are special in the fact that boy dolls, in general, are hard to find. Peter dolls are marked c Ideal Toy Corp / BE - 35 - 38 (on head) and c Ideal Toy Corp / W - 38 / Pat Pend (on body)

To my knowledge, all Peters were made to be walkers. Some, over the years, may have loosened up or may have been poorly restrung and no longer appear to be walkers.

Hair Colors: Sandy Blond, Auburn, Brown, Brown-Black (called Black by collectors - rare)

Eye Colors: Hazel (brown - green), Light green, Gold, or Golden Brown

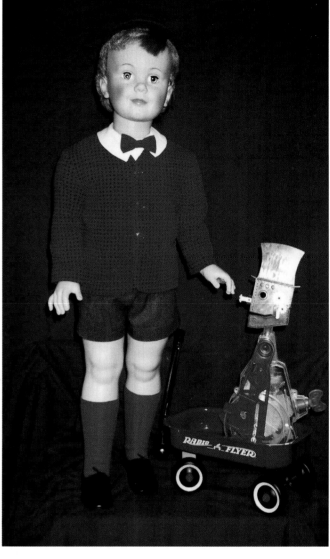

VALUES FOR PETER

VALUE - PETER PLAYPAL, IF MINT AND ORIGINAL - ABOUT $1200 TO $1300

Peter's original clothes basically consisted of blazers that were paired up with either short or long pants. He had three red blazers, one with a small blue stripe, one with a small blue check and one solid red corduroy called the Honey Mate outfit. His solid navy blue suit is considered hard to find as is his large plaid blazer/short pants outfit. *Courtesy of Kathy Ebey.*
(Please note: Original Peter Playpal shoes sell for about $50)

38 inch Peter Playpal (left) stands next to the very rare 36 inch Peter Playpal Salesman Doll (right), which is courtesy of the M. Hilliard Collection.

Peter Playpal, original and tagged, in a similar red jacket, but with a small, check-like print, instead of a stripe. *Courtesy of Laura Brink.*

Three Peters, including a hard-to-find black - haired Peter (center) are wearing the rare Blue Suit outfit. *Courtesy of the M. Hilliard Collection.*

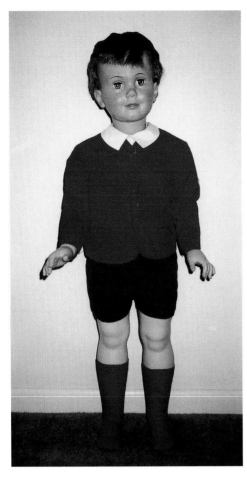

Peter Playpal in the rare Honey Mate outfit, which includes a red corduroy jacket and belt, red knit socks, a white shirt with a Peter pan collar, and navy blue shorts and cap. *Courtesy of Murray Hilliard.*

Above and below: Great close ups of Peter Playpal's original shoes and cap. *Photography by John Medeiros.*

Peter with brown/black hair or "Black Haired" Peter, and with original box. *Courtesy of Kathy Ebey.*

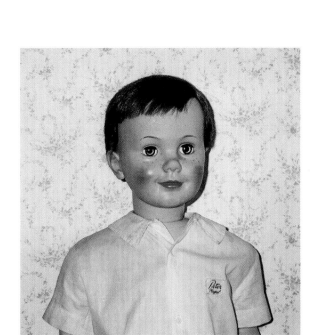

Peter Playpal, in original tagged shirt. *Courtesy Kathy Hosteller.*

INTERESTING FACT

The rare 36 inch Peter Playpal "Salesman" doll came about as a result of Playpal traveling salesmen complaining that the original 38 inch versions were too tall to fit in the trunks of their cars. Thus, as it is told, Ideal shrunk these salesmen's demos down by 2 inches. These shorter Peters are too rare to even come up with a value on!

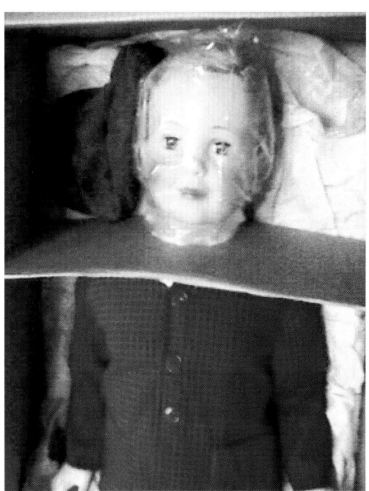

A rare glimpse of what Peter looked like, pristine and new on Christmas morning, 1960. *Courtesy of John Medeiros.* This degree of mint condition can add hundreds of dollars to the value of a Peter. *Photography by John Medeiros.*

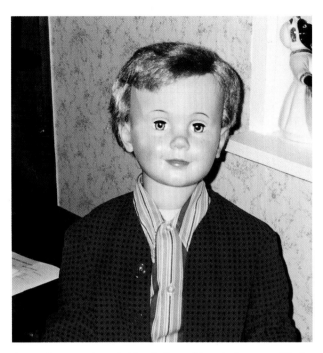

Great close-up of sandy blonde Peter Playpal. *Courtesy of Kathy Hosteller.*

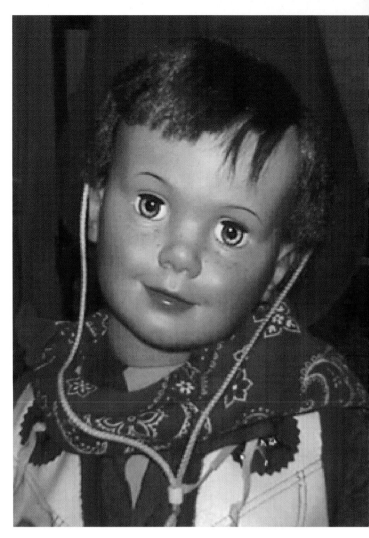

This is a darling close up of a redressed Peter. *Courtesy of Janet Porkrinchak.*

A beautiful example of a Peter Playpal with sandy blond hair and green eyes. He is redressed and stands next to a 1960 auburn Patti. *Courtesy of Tara Wood.*

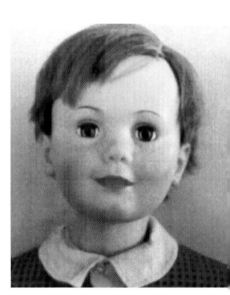

AUCTION NOTE: This mint Peter Playpal, in original outfit, with wrist tag but without box, sold for $1,235 in 1998.

PENNY PLAYPAL

PENNY PLAYPAL - FACTS AT A GLANCE

Penny was made for only one year, 1959. At 32 inches, she is the size of a 2 year old, and has a soft, rounded face that is endearing to anyone that sees her. She is marked "Ideal Doll / 32 - E - L" OR "B-32 Pat. Pend." (on her head) and "Ideal" in an oval on her back.

Hair Colors: Sandy Blond, Auburn, Brown, Dark Brown/Black
Eye Colors: Blue, Green, Brown

Photography by John Medeiros.

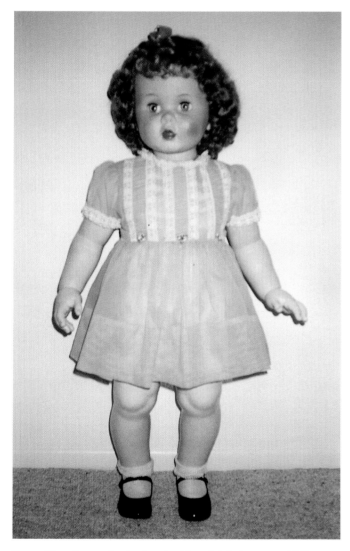

Penny Playpals can also be found with a slightly longer and fuller hairstyle. This beautiful example is completely original from top to bottom. *Courtesy of Laura Brink.*

A beautiful Penny Playpal, with original hair set. *Courtesy of Cheryl Kelly.*

VALUES FOR PENNY

VALUES - PENNY PLAYPAL, IF MINT AND ORIGINAL - ABOUT $400 TO $450

All original Penny and Suzy. *Courtesy of Tara Wood.*

Gorgeous example of an auburn Penny. *Courtesy of Kathy Hosteller!*

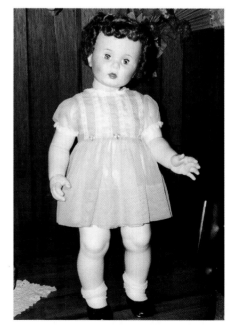
Breathtaking 32 inch Penny Playpal, in original dress. *Courtesy of Kathy Hosteller.*

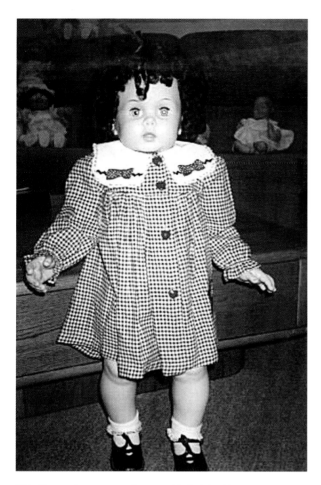

This Penny shows a most unusual hairstyle. Her owner says her hair is too nice and too tight to be a reset. *Courtesy of Lori Gabel.*
NO VALUE HAS BEEN ESTABLISHED FOR THIS DOLL

These brunette and auburn Penny Playpals show their original yellow and blue nylon dresses and original T-strap shoes. This dress also came in a light pink. *Courtesy John Sonnier.*

Auburn Penny Playpal. *Courtesy of Lori Gabel.*

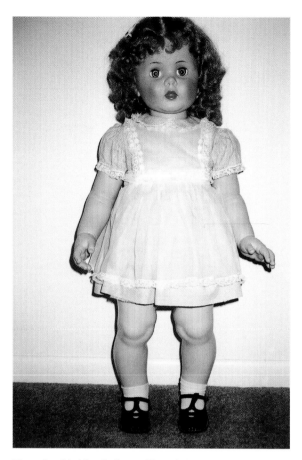
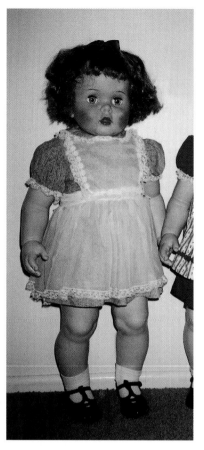

This adorable blonde Penny Playpal shows us the very rare Blue Checked Party Dress with the white organdy pinafore. *Courtesy of Murray Hilliard.*

Penny Playpal shows us a variation of the Blue Checked Party Dress, which is, instead, in a dainty red and white check. *Courtesy of Lori Gabel.*

SUZY

SUZY PLAYPAL - FACTS AT A GLANCE

An angelic baby doll that is the size of a 1 year old - 28 inches. In ads of the era, her name is spelled both Suzy and Suzie. She came in both straight and curly hair, with curly being the most common. Suzys are marked "Ideal Doll / O.E.B. - 28 - 55" OR "24 - 3" OR Ideal OB-28 (on head) and "Ideal Toy Corp B 28" OR "Ideal" in and oval on the back.

Hair Colors: Sandy Blonde, Light Brown, Auburn (rare), Brown/Black (rare) Suzy's hair was tightly curled. A rare version had straight hair.

Eye Colors: Blue, Brown, Green

VALUES FOR SUZY

VALUES - SUZY PLAYPAL, IF MINT AND ORIGINAL - ABOUT $300 TO $350

Black haired and blonde Suzy Playpals. *Courtesy of Kathy Hosteller.*

Rare black haired, curly haired Suzy. *Courtesy of Kathy Hosteller.*

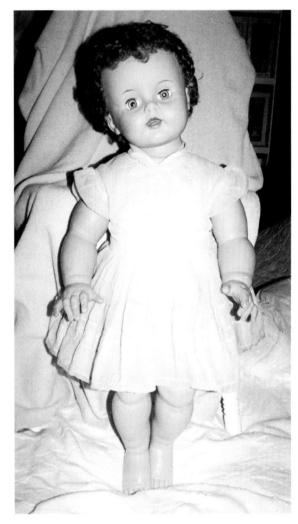

This is a great shot of Anita Holmes' brown haired Suzy, showing some detail of her chubby arms and legs.

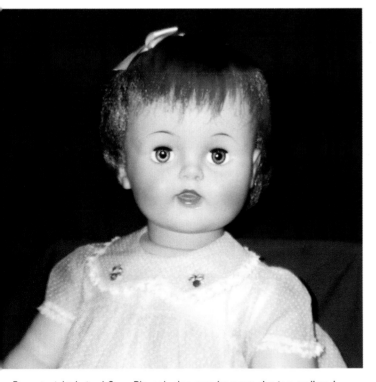

Rare straight haired Suzy Playpal, also rare because she is a redhead. *Courtesy John Sonnier.*

The difference between blonde and brown-haired Suzys is subtle, and has sometimes confused collectors when having to describe these dolls over the phone to buyers.

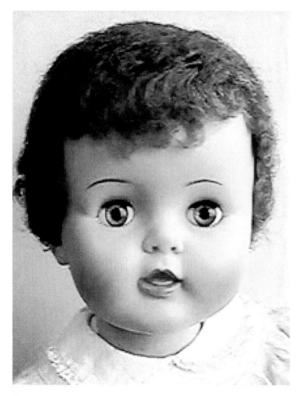

Example of the shorter hair length that some Suzys came with .
Courtesy of Kym Baker.

Suzy Playpal in original pink organdy dress. *Photography by John Medeiros.*

Black haired Suzy in hard to find original Bib Dress. *Courtesy of the Maxwell-Nieves Collection.*

Saucy Walker and Suzy are dressed in hard to find original outfits, blue and white cotton pinafore sets in this catalog advertisement.

TWINS BONNIE & JOHNNY PLAYPAL

TWINS BONNIE AND JOHNNY PLAYPAL FACTS AT A GLANCE

These dolls are exceptionally hard to find, with Johnny being the harder of the two. Bonnie has rooted hair and Johnny is the only Playpal doll with painted hair. They are 24 inches long, making them the size of 3 month old babies. They are marked as follows:

Johnny - Ideal Doll / BB - 24 - 3 (on head), Ideal (in oval) (on body)

Bonnie - Ideal Doll / OEB - 24 - 3 (on head), Ideal (in circle) 23 (on body)

Hair Colors: Bonnie - Blond, Dark Brown, Auburn; Johnny - Painted on Brown Hair
Eye Colors: Bonnie - Blue, Green; Johnny - Blue

Bonnie's original outfit was a blue and white checked dress, and Johnny's original outfit was a purple and white checked smock.

Photography by John Medeiros

VALUES FOR BONNIE & jOHNNY

VALUES FOR BONNIE OR JOHNNY PLAYPAL, IF MINT AND ORIGINAL - ABOUT $700.

Telling the difference between a Sandy Blond and an Auburn Bonnie can be confusing to collectors, because the blond Bonnies can have a reddish cast to their hair. Pictured is a Blond Bonnie. The Auburn Bonnie Playpals are a dark red. *Courtesy of Jon and Sue Hardgrove.*

Bonnie Playpal (top) shown with the dolls she is most frequently confused with, the Dryper babies (bottom). LOOK FOR A NURSER MOUTH TO IDENTIFY A DRYPER BABY. Bonnie's mouth is open/closed, but is not a nurser mouth. Dryper Babies are precious and adorable, but do not command Bonnie Playpal prices, so if you don't have all of the dolls' markings memorized, learn to tell the difference! *Both photos courtesy of Kathy Hosteller.*

INTERESTING FACT

Johnny Playpal, with his painted on brown hair, is harder to find than his twin sister Bonnie.

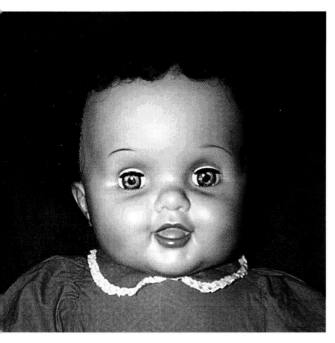

Bonnie Playpal with brilliant green eyes. *Courtesy of Delores Elliott.*

Redressed Johnny Playpal. *Courtesy of Kathy Hosteller.*

Johnny Playpal in replica of his original outfit. *Photography by John Medeiros.*

Left: Redressing Bonnie and Johnny is just too fun, as you can shop in any baby department! These two cute examples are courtesy of Tara Wood.

Above: Another example of Bonnie and Johnny. *Courtesy of Kathy Hosteller.*

CHAPTER 4
THE 36" SHIRLEY TEMPLES

Shirley Jane Temple was born in 1928. Her very first words were "Oui, mon cher" (French for "Yes, my dear") A few short years later, she became the first child to win an Academy Award. In time, she was voted number 18 of the top actresses before 1950 by the American Film Institute, won the hearts of probably everyone that drew breath, and made being adorable into a science.

A parade of Shirley Temple dolls were made over the decades, but the first life-size Shirley Temples were made by Ideal during the Playpal years. These are magnificent dolls and very respected by collectors.

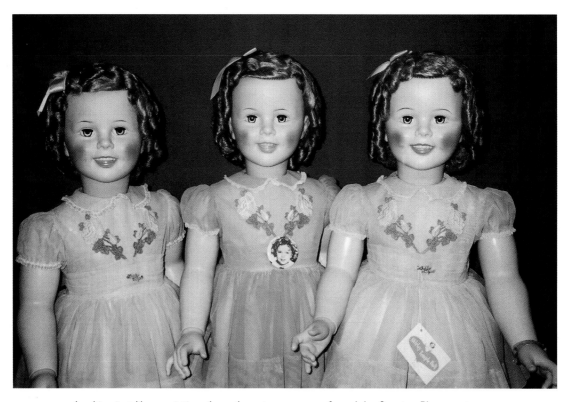

Looking just like a painting, these three treasures are from John Sonnier. Please note: the yellow dress this Shirley is wearing is the hardest to find of the three solid color (pink, blue, yellow) dresses in this style.

VALUES FOR SHIRLEY TEMPLE

VALUES - FOR 36 INCH SHIRLEY TEMPLE, IF MINT AND ORIGINAL - about $1400 to $1500

As you can see by this remarkable lineup owned by John Sonnier, there are many subtle, yet fascinating variations among Ideal's 36 inch Shirley Temples. The three nylon dresses with the "V" shaped flower appliques on the dress bodice are top row center. Please note that the dresses shown here are absolutely mint and when found in played with condition, they can be so many shades lighter that they may be very hard to recognize. The nylon dresses with the horizontal flower appliques (top row, left, and 3 in bottom row) can be found both solid colored and two-toned. All the dolls pictured have the jointed or "twist" wrists, except for the Shirley in the Heidi outfit. And, finally, look closely to see if you can detect subtle differences in the facial sculpting of these dolls.

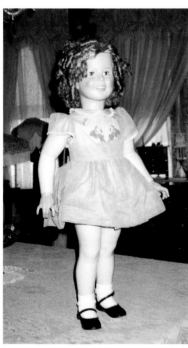

Example of original 36 inch Shirley Temple dress, in it's more frequently found condition of less than mint. The pastels of the three dress colors have frequently faded to a color that is several shades lighter than the original, and the flower appliques are frequently faded to an off-white. *Courtesy of Dee Della Vecchia.*

36 inch Shirley Temple with the rare fuller hairstyle. *Courtesy of John Sonnier.*

Hard to detect with the naked eye, but a cinch for an advanced collector like John Sonnier, this pair of Shirleys represent two different types of the "Playpal" Shirley Temple. The doll on the right is 1-1/2 inches shorter (34-1/2 inches), has a thinner face and her mouth is less wide and more open. We have seen similar variations in the faces of other vinyl dolls when a different vinyl has been used for the same face mold.

The Shirley on the left has a rare hairstyle, with thicker, longer curls and a lower side part. Eyes are noticeably darker and browner than the other Shirleys. *Courtesy of John Sonnier.*

Shirley, in another style of organdy dress. *Courtesy of Lori Gabel.*

Two original Shirley Temple nylon dresses with the horizontal flower appliques - the two-toned yellow and white and the solid pink. *Courtesy of John Sonnier.*

Extremely hard to find is the Shirley Temple "Heidi" outfit. A rare outfit like this can add $100 to $200 in value to the doll, if both the buyer and seller are aware of the fact. *Courtesy of John Sonnier.*

ODD - EYED SHIRLEY TEMPLE
The 36 inch Shirley Temples came with Hazel eyes. The doll shown here has one hazel (our right) and one green eye (our left). From the mint condition of both the doll and the eyes themselves, it can be determined that one eye did not change color due to moisture exposure, and that these are original factory eyes. This makes for an incredibly rare doll and a great conversation piece! *Courtesy of John Sonnier.*
Please note - The yellow dress this Shirley is wearing is the hardest to find of the three solid color (pink, blue, yellow) dresses in this style.

Rare variation of the 36 inch Shirley Temple, with prominent teeth. *Courtesy of Lee Epstein.*

This extraordinary Shirley, with the fuller face, is said to have come from the factory wearing this Patty Playpal dress, the Powder Blue Pin Stripe. *Courtesy of John Sonnier.*

Like Pattite, the 19 inch Shirley Temple, made by Ideal in the 1960s is a miniature version of her 36 inch Playpal counterpart. Additionally rare is the walker version of this 19 inch doll. She walks when guided in a heel-toe fashion. She is marked "Ideal Toy Corp. / ST-19-R" (on head) and "Ideal Toy Corp / G-19" (on back). She has a hard plastic body and legs with a vinyl head and arms. Her head and neck are molded as one piece. Shown here, she is wearing her original seersucker plaid dress with original shoes, socks, pin and wrist tag. This walker version also came dressed in the red and white dotted Swiss dress found on the standard 19 inch Shirley Temples. *Courtesy of John Sonnier.*
VALUE - ABOUT $1000

The standard 19 inch Shirley Temple doll (L), which is very hard to find, stands next to the very rare 19 inch Shirley Temple walker (R) both made by Ideal in the 1960s. Value of standard 19 inch Shirley Temple $800 to $900. *Courtesy of John Sonnier.*

THE 1984 36 INCH SHIRLEY TEMPLE
This limited edition of 10,000, stock #03309, was issued on July 15, 1984 by Dolls, Dreams and Love. She was designed by Hank Garfinkel. *Courtesy of Rebecca's Place in Oklahoma City, OK.* VALUE - ORIGINAL AND MIB: $250

19 and 36 inch Shirley Temples. *Courtesy of John Sonnier.*

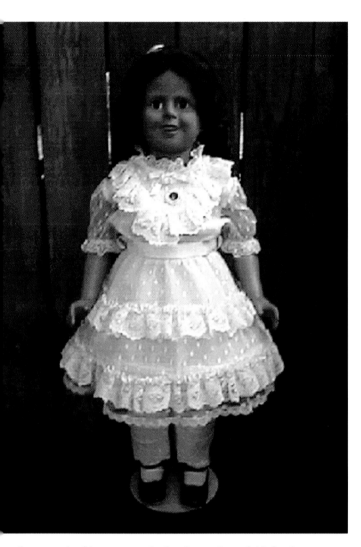

> INTERESTING FACT
>
> A 1985 reissue of the 36 inch Shirley Temple was produced by Dolls, Dreams and Love, a company owned by a former Ideal employee Hank Garfinkle. They are marked "17/1984 Mrs. Shirley Temple Black/ Dolls Dreams and Love."

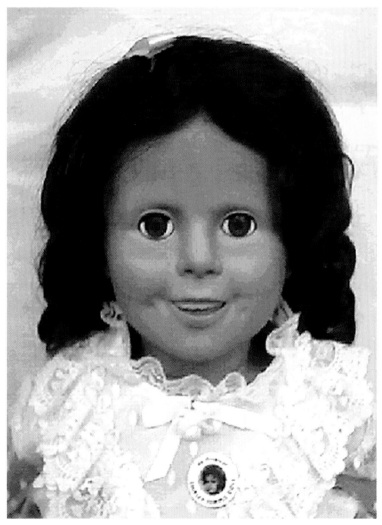

A rare and golden opportunity for the readers of this book to see the black 36 inch Shirley Temple doll, made in the mid 1980s. Only a very few of these were made and they were never put on the market. No value has been established for the black Shirley Temple dolls.

Another fun item for the collector of larger dolls is this life size Shirley Temple Paper Doll. Just like the Playpals and the 36 inch Shirley Temple dolls, this doll is sturdy enough to wear real clothes! Her paper costumes included in the box feature outfits from her movies "Captain January," "Wee Willie Winkie," "Rebecca of Sunnybrook Farm," "Little Miss Marker," and "Heidi." *Courtesy of the Treasure Mart in Puyallup, WA.*
VALUE - COMPLETE AND EXCELLENT - ABOUT $100)

CHAPTER 5
PLAYSETS, ORIGINAL TV & CATALOG ADS

THE PATTI PLAYPAL GAME by IDEAL
The Patti Playpal Game, which was "based on America's favorite doll" gives us fabulous photography of mint Patti s in their original clothing. The game itself is played in a manner similar to the famous Candyland, where the players advance to the next space that is the same color as the card drawn from the center pile. The object of the game is to move the tiny Patti Playpal paper dolls from "Start," across the board, to the Patti Playpal dollhouse in the center. Thank you so very much, Cindy Sabulis for these great photos!
VALUE - COMPLETE AND IN EXCELLENT CONDITION - ABOUT $100

The 1960-61 Gambles fall Catalog marketed a pink Rite Hite Steel Kitchen for Patty Playpal. The Patti Playpal doll, shown here in the ad in an Organdy Pinafore dress, sold through this catalog for $22.95, while the entire kitchen sold for about $30.

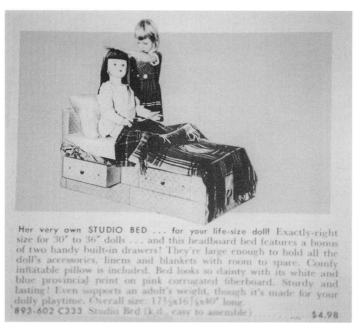

A pink and blue corrugated cardboard bed was sold for Playpal sized dolls. The bed included storage drawers and an inflatable pillow.

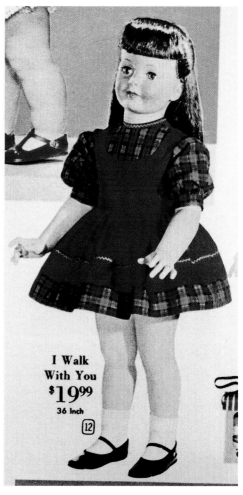

The 1961 Montgomery Wards Christmas catalog shows a brown–haired Patti in a variation of the Plaid School Dress, a darling red and green Christmas-style dress with ric rac trim.
VALUE - DRESS ONLY - WITHOUT UNDERWEAR OR SOCKS, IN EXCELLENT CONDITION - ABOUT $125

INTERESTING FACT

The cover of the Patti Playpal Board Game shows Patti in a ricrac pinafore dress that has a yellow floral/stripe cotton underdress.

A 1965 magazine ad for Grant's Department store spotlights companion dolls.

Little girls could get matching dresses for their companion dolls through the 1961 Wards Christmas catalog!

The Sears 1961 Christmas catalog shows Patti Playpal available in "assorted style outfit." These types of catalog finds make documenting original clothing such a delightful challenge! Shown here in a checked smock top worn over a dress, she sold for a mere $18.88! Oh, to have a time machine and travel back to 1961 for one frenzied day of shopping! *Courtesy of Cindy Sabulis.*
VALUE - DRESS ONLY - WITHOUT UNDERWEAR OR SOCKS, IN EXCELLENT CONDITION - ABOUT $100

ADORABLE LIFELIKE DOLL

Your littlest lady will be ecstatic with a Barbara Jo Playmate! She wears regular 2-yr. old's dress, lingerie, shoes, socks... eyes, with lovely lashes, close...she's washable and fully jointed...her hairdo can be washed, combed and brushed $9.98

This 1960 Rexall ad shows "Barbara Jo Playmate" for $9.98.

CHILD-SIZE SISSY DOLL... FULL 36" TALL!
Brace yourself for a headlong hug and shrill shrieks of joy when you present this life-size doll playmate to a little girl! Sissy's full-jointed body is appealingly molded in lightweight plastic—easy for little arms to carry. Her pretty head with halo of rooted curls is realistic vinyl. Sissy stands alone, and sits down for a doll tea party. Her long-lashed eyes close at bedtime. She wears a perky blue-and-white-striped dress topped by a fresh, sheer, white apron that's trimmed with lace. Panties, shoes and socks.
87-27-54 B998 36" Sissy Doll......$14.98

36 inch Sissy doll sold for $14.98 in a Chatty Cathy style dress.

A big thank you to Earthly Remains of Mankato, MN for help on these original catalog pages.

EARLY TELEVISION ADS

Saturday morning television commercials kindled and captivated the hearts and minds of America's 1960s children. Like deer staring into headlights, more than a few of us were mesmerized by the images of the Playpal dolls and the child actors they seemed to dwarf. After watching weeks of these commercials, the dolls seem to take on exaggerated prominence. When you see a doll under the Christmas tree that you've seen over and over again on television, it is as though you are receiving a celebrity into your home!

The original 1960s Patti Playpal commercial read as follows:

> *Look who's come to your house to stay!*
> *A pal you can play with all day long!*
> *She's Patti Playpal...*
> *A big doll - big as a three year old!*
> *And Patti is real as life!*
>
> *Patti's beautiful hair can be brushed, braided and put into a ponytail. And Patti Playpal, big as she is - is light and easy to carry.*
>
> *You can dress her in your own clothes, because Patti Playpal is almost your size! She can wear your pajamas and stay overnight.*
>
> *Yes, Patti's a wonderful friend! She's just one of the Ideal Playpal family. Starting at $12, there's Johnny Playpal, Suzy Playpal, and Penny Playpal too!*
>
> *Meet Patti and all her Playpal family at your favorite store.*
>
> *She's a wonderful doll - she's Ideal!*

All photos of original television commercials are made possible by "Video Doll Shop" courtesy of Video Resources of N.Y. and Ira Gallen.

CHAPTER 6
PORTRAIT GALLERY

CHAPTER 7
OTHER LARGE DOLLS OF THE ERA

THE LORI MARTIN (or VELVET BROWN) DOLL

Every Sunday night in the early 60s, NBC-TV and Rexall presented MGM-TV's "National Velvet." Lori Martin was the beautiful young actress that played the lead role of Velvet Brown. Her image appeared in TV Guide, on coloring books and paper doll sets. The Lori Martin or Velvet Brown doll made in her likeness is a sought after and pricey doll on today's collector's market.

Lori Martin dolls were made around 1961. They came in 30, 36, 38 and 42 inch sizes. Her original outfits consisted of plaid shirts, jeans, boots and cowboy hats and one dress. All Lori Martin dolls are marked Metro Goldwyn Mayer and Ideal Toy Corp.

VALUE - FOR LORI MARTIN DOLL, 30, 36 AND 38 INCH SIZES, IF MINT AND ORIGINAL, ABOUT $1300
42 INCH LORI MARTIN DOLL, IF MINT AND ORIGINAL - ABOUT $1700

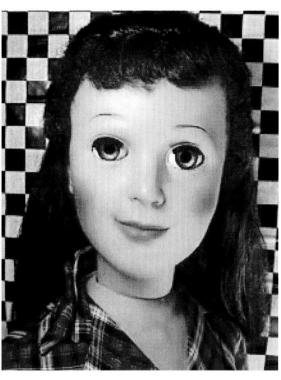

Photography by John Medeiros.

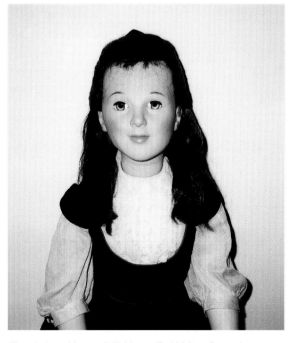

42 inch Lori Martin doll (also called Velvet Brown), mint and original. This is the very rarest of sizes for the Lori Martin doll. *Courtesy of Murray Hilliard.*

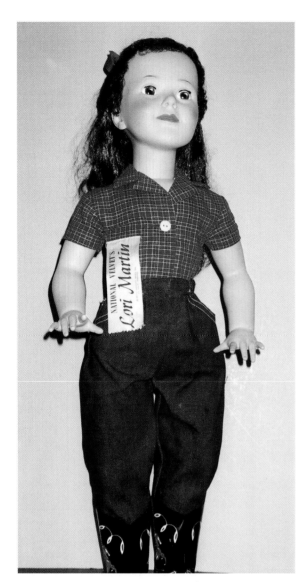

30 inch Lori Martin doll. *Courtesy of Shari Wentz.*

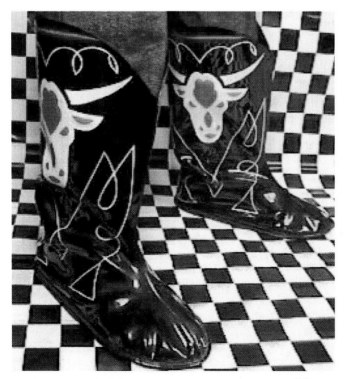

Close up of the detail on the riding boots of the Lori Martin (or Velvet Brown) doll. *Photography by John Medeiros.*

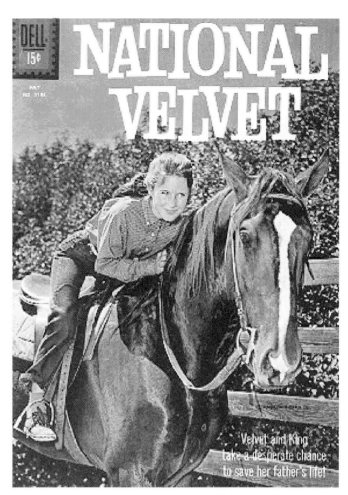

First National Velvet comic book with Lori Martin and King on cover. *Courtesy of Charles Neatfunstuff.*
VALUE - VERY GOOD: $10

An interesting item for collectors of the Lori Martin or "National Velvet" doll, the 1961 "National Velvet" Whitman coloring book, stock # 36563123. There was also a 1963 version of this coloring book and both are valued at between $10 and $15.

Another fun collectible to display your Lori Martin dolls with - the 1961 Whitman "National Velvet" paper dolls, with the very beautiful Lori Martin on the cover. Original price, 29 cents. VALUE today - about $10 in good condition, and about $25 uncut. *Courtesy of Ann Consler.*

DADDY'S GIRL

Daddy's Girl by Ideal, a hard to find doll, came in two sizes, 42 and 38 inches. The 38 inch size is extremely rare, and is valued at about $300 more than a 42 inch Daddy's Girl in the same condition. Daddy's Girls came in three hair colors - blonde, brunette and auburn, with auburn being the hardest to find.

VALUES FOR DADDY'S GIRL DOLLS:
MINT AND ORIGINAL, 42 INCH BRUNETTE OR STANDARD BLOND: $1400
SAME, BUT AUBURN: $1800
38 INCH DADDY'S GIRL - (ADD $300)

This is an all-original 38 inch Daddy's Girl. *Courtesy of John Sonnier.*

The very rare 38 inch Daddy's Girl (left) wears a replica party dress with her original shoes. To the right is a 42 inch Daddy's Girl in her original tagged dress. *Courtesy of Murray Hilliard.*
VALUE - 38 INCH BRUNETTE, MINT WITH REPLICA DRESS: $1600
42 INCH BRUNETTE, ORIGINAL, $1400

Very hard to find Auburn Daddy's Girl in rare variation of the plaid dress, with a rounded collar. May have been a prototype.
VALUE, IF MINT: $2200 (ADDED VALUE INCLUDES BOTH HARD TO FIND HAIR COLOR AND DRESS VARIATION)

Daddy's Girl in replicas of her original outfits, sewn by Barb Byrnes. *Photography by John Medeiros.*
VALUES - BLONDS, MINT, IN REPLICA OUTFITS: $1300 EACH

Unusual variation of Blond Daddy's Girl, with a lighter shade of blond than the standard, in original dress. *Courtesy of Kathy Ebey.*
VALUE: $1600

42 inch Daddy's Girl next to a sandy haired Peter, both redressed. *Courtesy of Kathy Hosteller.*
VALUE - REDRESSED BLOND DADDY'S GIRL, GOOD CONDITION - ABOUT $900

SAUCY WALKER

Saucy Walker, made by Ideal in 1960, is an adorable toddler doll that Playpal collectors seem to be fond of. Much different in look than her 1950s counterpart, she is a winning combination of chubby cheeks and a smirk-ish grin.

Saucy came in two sizes – 28 inch and the 32 inch size which is sometimes referred to as "Super Saucy Walker" or "Playpal Saucy Walker" by collectors. She came originally in either a blue print dress with a white pinafore (32 inch) or a solid red dress with a red print pinafore (28 inch).

The 32 inch size is marked "BYE – 3235" and the 28 inch size is marked "T28 X – 60." Both dolls are marked Ideal.
VALUES, MINT
28 INCH - $175
32 INCH - $200

This is the Super Saucy Walker, made by Ideal in 1960. This doll is frequently referred to as the "Playpal Saucy Walker" by collectors. She is redressed. *Courtesy of Nita Moulton.*

Photo from original Saucy Walker commercial. All photos of original television commercials are made possible by "Video Doll Shop" courtesy of Video Resources of N.Y. and Ira Gallen.

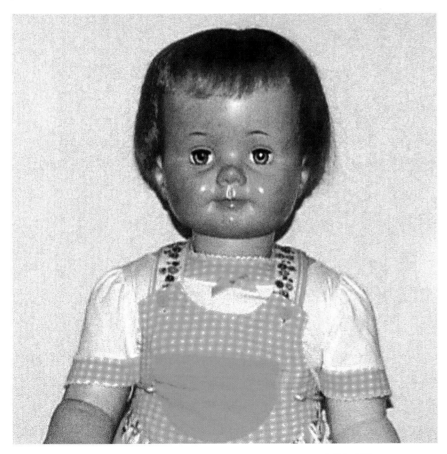

Beautiful close up of the 32 inch (or Super) Saucy Walker, marked "Ideal Toy Corp/ BYE – 32 – 35" *Courtesy of Debi Toussaint-Edgar; photo by Steve Edgar.*

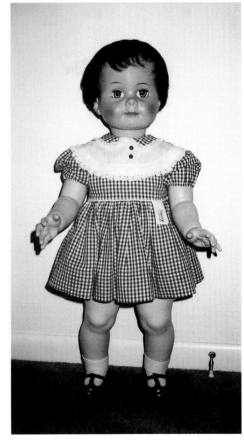

Extremely rare Red Checked tagged Saucy Walker dress. *Courtesy of the M. Hilliard Collection.*

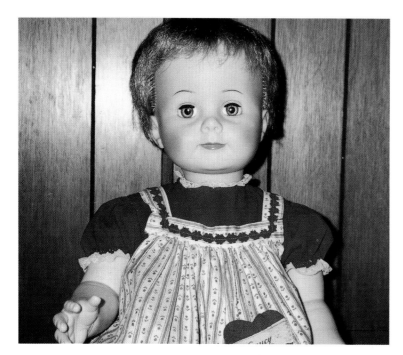
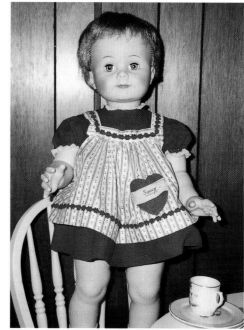

28 inch Saucy Walker in original dress. *Courtesy Kathy Hosteller.*

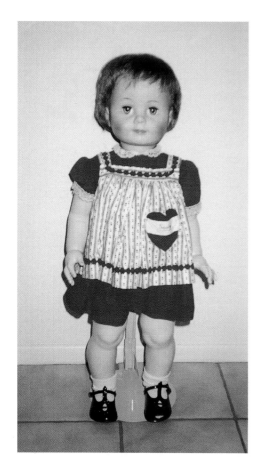

Slight variation of 28 inch Saucy Walker's original dress, with a shorter pinafore. *Courtesy of Laura Brink.*

Both the 28 and 32 ("Super" or "Playpal") inch Saucy Walker dolls came in this bright blue print cotton dress with white cotton pinafore with "Saucy Walker" label. *Courtesy of John Sonnier.*

Example of odd-eyed Saucy Walker. I believe these odd eyes to be due to some kind of moisture exposure.

INTERESTING FACT

In the 1960 Sears Wish Book, Saucy Walker was called "Chubby 2 Year Old," and sold for $21.88 for the 32 inch version and $18.88 for the 28 inch version.

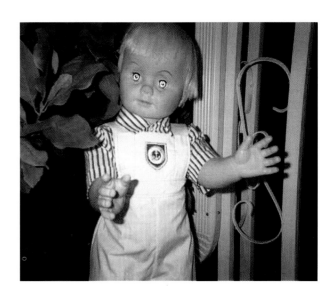

SAUCY BOY?

Even though I cannot identify him, I wanted to include this little guy in the book, because I know he will be of interest to the readers. This unmarked boy doll is a dead ringer for Saucy Walker's twin brother. He is a walker, has the same vinyl type as the Playpal dolls, and is identical to the Saucy Walker doll, but because he is not marked, one can only conjecture where he came from. This could be a doll that was never put on the market. *Courtesy of Hazel Schwartz.*
NO VALUE HAS BEEN ESTABLISHED FOR THIS DOLL.

MISS IDEAL

Miss Ideal is jointed at wrists, above knees and at waist. Came in both 25 and 30 inch sizes. Miss Ideals were a dark blond - a rare version came with platinum blond hair. Marked c Ideal Toy Corp / SP - 30 - 6 (on head) and c Ideal Toy Corp / G - 30 - 6 (on back) Miss Ideal dolls came with pretend "perm" kits that included curlers, comb and waving lotion.
VALUES - MISS IDEAL, IF MINT AND ORIGINAL - $150

Some of Miss Ideal's Original Outfits:
1) Town and Country Outfit - yellow and white checked sleeveless cotton dress with black velvet jacket, straw hat.
2) Campus Outfit - pink full-skirted dress with embroidered nylon overlay and red trim.
3) Gold Capri Set - Gold cotton capri pants and smock.
4) Pink Gingham Capri Set - pink gingham smock with flower applique and dark capri pants. (pictured)
5) Blue Gingham Capri Set - blue gingham midriff top with flower applique and black corduroy capri pants. (pictured)
6) Black and White Check - black and white taffeta check skirt, black velvet jacket. (pictured)
7) Pink Cotton - pink jumper-style dress with black and white checked blouse. (pictured)
8) Kerchief Dress - (Also called Town and Country Dress in catalogs) - blue full skirted dress with print overskirt apron and matching triangle scarf. (pictured). Similar dress and scarf came in green. (pictured)

Miss Ideal dolls came with pretend "perm" kits that included curlers, comb, and waving lotion

Beautiful close up of Miss Ideal. *Courtesy of dorlin-dolls.com.*

Same outfit with original flower applique on a brunette Miss Ideal *Courtesy of Kathy Ebey.*

Miss Ideal in original outfit, with replaced flower. *Photography by John Medeiros.*

Miss Ideal as she appeared in the 1961 Wards Christmas Catalog.

Miss Ideal's Blue Kerchief Dress, also called one of the Town and Country dresses.

Miss Ideal's Green Kerchief Dress. *Courtesy of Kathy Ebey.*

The 1961 Wards Christmas catalog shows 2 original outfits for Miss Ideal.

BYE BYE BABY

Adorably redressed Miss Ideal. *Courtesy of Nancy Sanderson.*

Bye Bye Baby was one of the life-sized baby dolls that Ideal made during the 1960s. *Courtesy of Kathy Hosteller.*
VALUE - IF MINT AND ORIGINAL - ABOUT $450.
REDRESSED - ABOUT $250 TO $300.

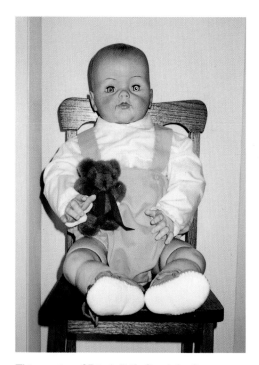

This precious 27 inch "Life-Size Infant" was Horsman's answer to Bye Bye Baby. This doll is multi-jointed and very hard to find. Value is about $100. *Courtesy of Murray Hilliard.*

Ideal's Bye Bye Baby, redressed. *Courtesy of Lori Gabel.*

JOANIE

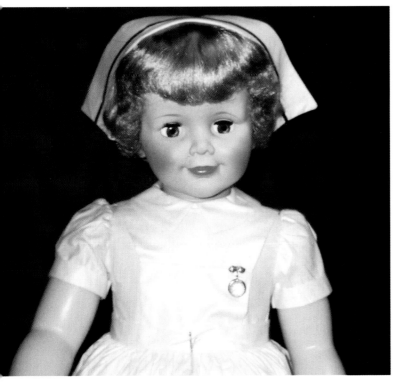

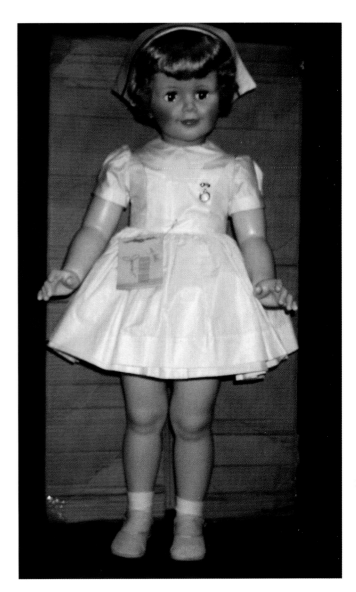

Madame Alexander's Joanie, with a straight hair style. She is shown here with her original box, her "watch" pin and her memory book/tag that is pinned to her apron bodice. She has a metal rod-walking mechanism that runs through her body and strung arms. Joanie was made in 1960 in a white nurse's dress and in 1961 in a colored dress with a white apron and cap. She is 36 inches tall and is marked "ALEXANDER 1950" on her neck. In this mint and original condition with box, she is worth about $800. *Courtesy of John Sonnier.*

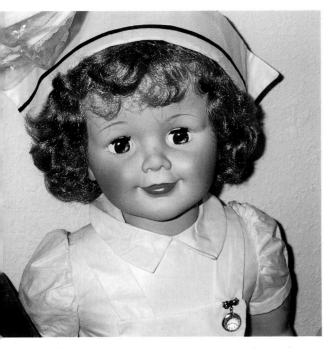

36 inch Madame Alexander Joanie, with a curlier hair style, original and mint. VALUE about $650. *Courtesy Kathy Hosteller.*

INTERESTING FACT

The Joanie doll sold in the Sears 1960 with an all steel nurse's cart that contained over 30 hospital play items such as forceps, stethoscope, and medical charts, for $28.88.

BETSY McCALL

VALUES FOR THE 36 INCH BETSY McCALL DOLL, IF MINT AND ORIGINAL - ABOUT $350

LINDA OR SANDY MC CALL, IF MINT AND ORIGINAL - ABOUT $250 TO $300

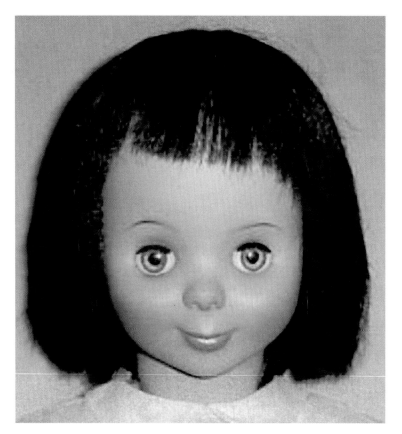

36" Betsy McCall marked "McCall Corp. 1959." *Courtesy of Mary Aikins.*

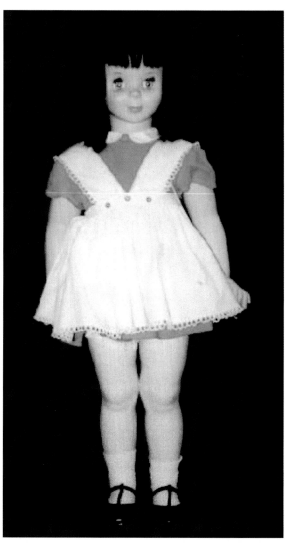

36" Betsy McCall in original dress. This dress also came in bright red with a white pinafore. *Courtesy of Annette Nott.*

This close up of Linda shows that distinctive McCall up-turned nose! This feature makes the McCalls easy to pick out in a crowd of companion dolls. *Photography by John Medeiros.*

35 inch Linda McCall, original with tag. *Courtesy Kathy Hosteller.*

Sandy McCall. *Photographed by and courtesy of John Medeiros.*

Two Sandy McCall dolls, in original clothing. *Courtesy of Lori Gabel.* VALUE - MINT AND ORIGINAL - EACH ABOUT $250-300

MARY JANE

Although Effanbee's Mary Jane is not a Playpal, she has managed to turn the head of many a Playpal collector. As you would expect from a company like Effanbee, her overall quality is above and beyond the great majority of Playpal look-alikes.

She is a 32 inch vinyl walker with flirty eyes. She was produced by Effanbee from 1959 to 1962, and originally sold for about $20. The hair colors that I have uncovered are light blonde, dark blonde, brown and black. She is marked "Effanbee Mary Jane" on her head. (In a perfect world, all dolls would be marked this clearly!)

Her original outfits are:
1) Pink and white striped cotton dress (shown)
2) White cotton nurse's dress with nurse's cap and striped apron.
3) Red organdy dress with white cotton embroidered pinafore.
4) Cobbler apron (large pockets that border the hem of the apron) worn over a cotton print dress – unsure of color.

She was also sold in combination with a baby doll in a set called Mary Jane and Lil Darlin. Set included bassinet with pink and white striped bedding to match Mary Jane's dress. Another set was sold with Mary Jane wearing her Nurse's Dress.

In 1963 the name "Mary Jane" was used by Effanbee as a 13 inch toddler doll.

If you dearly love these child – size dolls, seek out a Mary Jane - she'll probably win you over.

VALUE: IN MINT CONDITION, DOLL ONLY, NO ORIGINAL OUTFIT: $150

Effanbee's Mary Jane in original dress, shoes and parasol not original.

Some Playpal collectors have a weak spot for Mary Jane.

Auburn Mary Jane. *Courtesy of Bill Stafford.*

Effanbee's Mary Jane and Lil Darlin Bassinet set, in original catalog ad.

PRINCESS PEGGY

Princess Peggy, Horsman's answer to Patti Playpal, was marketed from 1960 to 1966. During those years, she had quite a variety of hair colors and styles – from bobs to ponytails – and at least 20 different original outfits. She was also made in an African American version and was a walker. Her drawn up little rosebud mouth and doe eyes give her a Pollyanna type of innocence. In 1967, Horsman replaced her with a companion – sized doll of lesser quality named Cindy.

Princess Peggy is marked "Horsman – 1959" on her head.

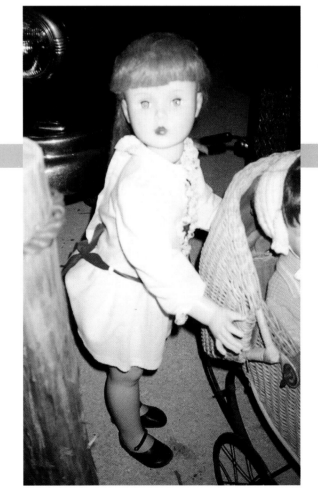

The doll pictured is courtesy of Collector's Showcase in Sturgeon Bay, Wisconsin. This doll, in Very Good condition and redressed, has a value of about $100.

Another example of a Princess Peggy, who came in a variety of hair colors and styles. *Courtesy of Manuel and Mary Cotta.*
VALUE REDRESSED AND EXCELLENT - ABOUT $100

Horsman's Princess Peggy in one of her many original outfits and hairstyle variations. Ward's 1961 Christmas Catalog.

This gorgeous Princess Peggy was a real competitor to Patti Playpal at only $10.97! A Patti Playpal sold in this very same catalog for $19.99. Ward's 1961 Christmas Catalog.

Horsman's Walk-A-Bye Doll set, made in 1960. The Princess Peggy walker is 36 inches tall and the baby doll in the pram is a 25 inch drink and wet. The 36 inch doll was attached with metal clips at the left arm and leg and could walk along with the carriage as a big sister. Ward's 1961 Christmas Catalog.
VALUE - ENTIRE SET, IF MINT AND COMPLETE: $300: $350.

30" LINDA WILLIAMS DOLL

Not many people know this about me, but at the age of 9, I was the official President of the Kedvale Avenue, Skokie, Illinois, Angela Cartwright Teen Beat Fan Club, which covered almost a 2 block area. With all of those titles, I should have been decorated like a four star general, but actually, I was the only one with the 85 cents needed to send away for the official fan club card with the official stamp and the almost autographed photo. Angela Cartwright was the actress, just about my age, who played the character of Linda Williams in the hit TV sit com "Make Room for Daddy." This show, which starred Danny Thomas, ran from the late 50s to the mid-60s and portrayed a nightclub singer and his family of a wife and three children living in New York City.

A doll was made by the Natural Doll Company to represent Angela (or her character, Linda Williams). A 14 inch model, which was available as a mail-in premium through General Foods is, today, fairly easy to find. The 30 inch model, however, which was probably inspired by Patti Playpal's popularity, is an very hard doll to find. Marked "Linda Williams" on head and in her original dress.

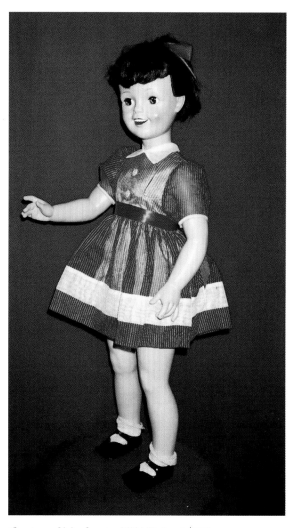

Courtesy of John Sonnier. VALUE about $450

Courtesy of Lori Gabel.

BUFFY

It has been written that the Playpal-sized Buffy doll was manufactured by Horsman, but I cannot determine this to be true. She is marked "36 - 2" on her head and was made to resemble the actress Anissa Jones, who played the character of Buffy on the 1960s television show, "Family Affair." Buffy was one in a set of boy/girl twins, Buffy and Jody, and was frequently seen with her doll sidekick, Mrs. Beasley.

This Buffy doll has a pull string and a grill and says things like Carry me please - I'm a big girl - and Please change my dress. Because of the TV tie-in, this doll is very sought after by collectors.

This same doll was also sold under the name "Talking Tandy."

Value - without original dress, which is how it is usually found, and in good condition, about $100. If mint, without dress, about $175.

LITTLE MISS ECHO

Little Miss Echo was made by American Character in 1962 and was 30 inches tall. Her promotions read "She repeats your words and songs in your own voice." Her body cavity housed a tape recorder which was activated by a knob in the torso area. 29 inch beautiful green or blue sleep eyes.

VALUE - Mint and Original, $75; w/ box, $100

Little Miss Echo and her original box,. Courtesy of John Sonnier.

BETSY WETSY

Betsy Wetsy is one of the many life-size baby dolls that Ideal produced during the Playpal era. This beauty was made in 1965 and is marked "c Ideal Toy Corp OBW – 20 – F" (on her head) and "BW – 20" (on her back)

This doll, in excellent condition and in original clothing is worth about $45.

IDEAL KISSY

22 inch 1963 Ideal Kissy. *Courtesy of dorlin-dolls.com.* VALUE Original and mint about $100.

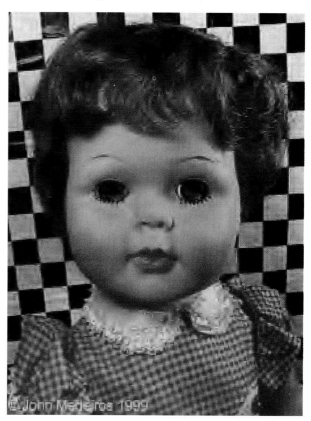

This mint in box Canadian Kissy, sold at auction for just over $175!! The Canadian dolls can be very intriguing to collectors, because of their subtle variations. As, you can see, this dolls face is slightly different from her American sisters. She was manufactured by Reliable of Canada. *Photography by John Medeiros.*

1964 MARY POPPINS BY HORSMAN

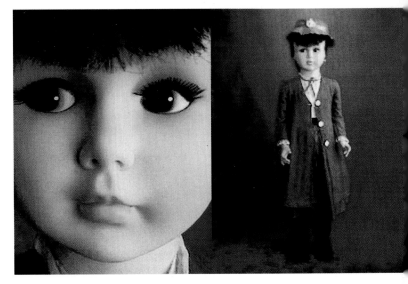

This original Mary Poppins, who is missing her umbrella, is marked "Horsman Doll #4" on her head. She is 36" tall, and because of the movie tie-in, she is valued in this mint and all original condition at $250. *Courtesy of dorlin-dolls.com.*

Beautifully redressed Mary Poppins. *Courtesy of Nancy Sanderson.*

SNOW WHITE

The Snow White doll, pictured with other Snow White memorabilia. Manufacturer unknown. *Courtesy of David M. Cobb Auction Service - Johnstown, Ohio.*
VALUE $250

LITTLE ORPHAN ANNIE

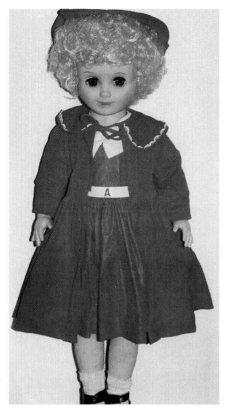

The 25 inch Little Orphan Annie doll, was issued by the Chicago Tribune/Daily news in 1958. It seems that very few of these dolls were made. She has a soft vinyl body and a rigid vinyl face. *Photo reprinted from* Doll World Omnibook, *published by House of White Birches..*
VALUE, IN VERY GOOD CONDITION AND IN ORIGINAL OUTFIT - ABOUT $400.

ARRANBEE'S MY ANGEL & VOGUE'S LIFE-SIZE GINNY

30" MY ANGEL TODDLER is dressed for a stroll in chic duxkin coat and beret . . . carries a matching bag! This fully-jointed toddler can *really walk* when you hold her soft little hand. (Arms, hands and head are all flexible vinyl with realistic "skin" texture). Her durable polyethylene body will withstand plenty of play-time scrambles without a mar. Long-lashed eyes close; saucy side-part bob is rooted. My Angel's white coat with bright brass buttons is lined in red taffeta to match her dress. Panties, socks and suede slippers complete her outfit.
87-27-62 B1265 30" My Angel Toddler **$17.98**

The 30 inch My Angel Toddler doll was a walker and sold for $17.98. This doll was made in the late 50s by Arranbee. She is unmarked. The mold for this doll was later sold to Vogue, and used for the 1960 life size Ginny, who came with an identically dressed 8 inch Ginny doll.
VALUE: IN MINT CONDITION, DOLL ONLY, NO ORIGINAL OUTFIT - $100

Two darling 27 inch Goody two Shoes, both originally dressed. *Courtesy of Murray Hilliard.*
VALUE Mint and Original about $75.

UNEEDA COMPANION DOLLS

BETTY BIG GIRL

Another favorite with the collectors of larger dolls is 32 inch Betty Big Girl, made by Ideal in 1969 and 1970. She is marked with "Ideal." Shown in her original outfit, she both walks and talks. Crissy collectors will notice that her face has a definite resemblance to Velvet's. *Courtesy of M. H. Hilliard Collection.*
VALUE - ORIG & EXCELLENT - ABOUT $80

Uneeda 30 Doll. These 36 inch companion dolls, circa 1960s, are marked "Uneeda 30" on their necks and are typical of the era. Uneeda 30 dolls are found frequently with large and striking turquoise eyes. *Courtesy of Sally Baldwin.*
Value is about $40.

Uneeda also jumped on the band wagon when companion dolls became popular. This doll, whose Mommy saw fit to give a haircut to, is marked "Uneeda 8." In this condition, she is worth about $30. In good condition, she is worth about $50. She is valued at a bit more than some of the companion dolls in this chapter because she is marked with her company's name, and as a general rule, collectors will pay more for this. *Courtesy of Dani Schade Vintagfind @aol.com.*

Uneeda's 1960 Pollyanna doll, inspired by the movie "Pollyanna," which starred Hayley Mills. At 32 inches tall, this same doll is often confused with the 1961 Uneeda doll known as Princess from the film "Babes In Toyland," and with the Uneeda doll with pinkish hair known as Fairy Princess. As you can see by this photo and the Princess and Fairy Princess dolls that follow, both the mouth molding and markings of Pollyanna are different from the two Princess dolls. Pollyanna is marked "Walt Disney / Prods / mfg by Uneeda / 5 / N F" (on head) Originally priced at about $15, she is today worth about $55, when found in excellent condition and in original outfit. *Courtesy of Clara Sue Arsdorff.*

An example of the "Princess" doll from the film "Babes In Toyland," made by Uneeda. Her marks are "c Uneeda Doll Co Inc." and "4" in circle. She is rewigged and redressed, but still a cutie! Thank you, Dani Schade! Her value in this condition is about $20. In good condition, with original hair, she is worth about $40

Here is the "Fairy Princess" doll by Uneeda. An adorable doll, she is marked "c Uneeda Doll Co. Inc." and "26" in circle. Redressed and in good condition she is worth about $40. The Fairy Princess doll is the same doll as Uneeda's Princess, but with silvery blond or pinkish hair. *Courtesy of Dani Schade.*

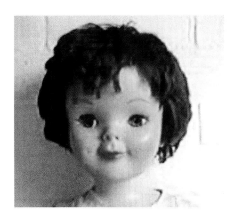

This same doll was later manufactured, unmarked, as a brunette. Value, in good condition, is about $30. *Courtesy of Kathy A. Peeler.*

African American Playpal look-alike with beautiful coloring. She is marked "Uneeda 3" on her head. Black dolls, as a general rule, are harder to find and have higher values. This doll, in her wonderful condition, is worth about $60. *Courtesy of Debbie Garrett.*

Uneeda's Fairy Princess doll, as she first appeared.

S 13

These "S 13" dolls (marked S 13 on head) are found fairly frequently at flea markets and antique malls. They are of lesser quality vinyl than most companion dolls, with visible seam lines. This doll, in played with condition and redressed has a value of about $25.

C DOLL

Another frequently found doll, while out on Playpal safari is the mysterious "C Doll" (marked "C" on her head and "35-5" on her back). They so obviously have the Playpal limbs – it was assumed that some other company obtained the original Ideal molds for her manufacture, but according to Neil Estern, Patti Playpal's sculptor, the limbs for the Playpal dolls were just copied outright by other doll companies. At 34 inches, she is a good quality doll, and has a value, in good condition and redressed of about $40.

Another beautiful example of a "C" doll, with long brown hair in a classic Patti Playpal style, marked "C" on head and "35 - 3" on back. She is redressed and in very good condition. *Courtesy of Mary Williams.*
Value: $40.

This "C" doll shows quite a resemblance to Patti Playpal. Some of these dolls have a "C" on their necks that resemble an "O." *Courtesy of Sherrie McCloughen.* VALUE - IN GOOD CONDITION, ABOUT $40.

AUCTION NOTE: This "C" doll was recently auctioned and the highest bid was $23.50.

J-CEY

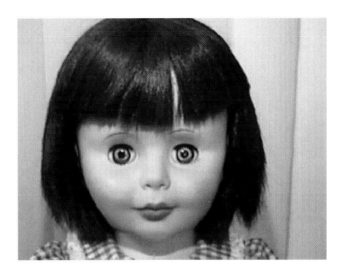

This 27 inch J-Cey doll, redressed in modern Playpal clothes, bears a facial resemblance to the "C" dolls. This doll, without the 1981 Playpal dress, is valued at about $25. *Courtesy of McCorrea.*

1966 LORRIE COMPANION DOLL

The Lorrie doll company, which produced a variety of baby and toddler dolls in the 1960s, also issued their own companion - sized doll during this era. At 36 inches, she is marked "1966 Lorrie Doll," and her most outstanding feature is a very pretty shade of blue eyes. Beacause the Lorrie brand is not notably collectible, these Lorrie companion dolls, in good condition, are worth about $40. *Courtesy of Robin Reece.*

"3" DOLLS

Using body and limb molds that are not like the original Playpal molds, these "3" dolls are frequently found by collectors of companion dolls. Because they are unlike the Playpal dolls in their sculpting, they are valued at a little less than other unnamed companion dolls. This doll, rewigged and not in original clothing, is worth about $20.

"5" and "6" DOLLS

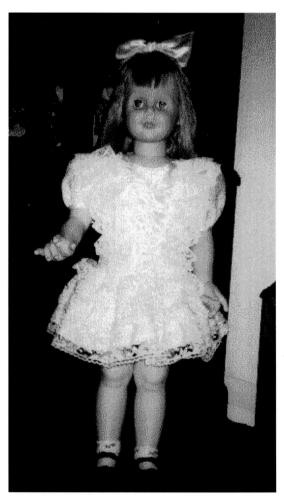

Using the same body mold as Patti Playpal, dolls marked "5" and "6" pop up from time to time. These dolls, when found in very good condition are worth about $40. *Courtesy of Judie Smith.*

"8" DOLLS

These 26 inch "8" dolls look a bit like Effanbee's Mary Jane, but are unmarked except for an "8" on the back of their necks. In excellent condition their value is about $30. *Courtesy of Morgan's Antiques – Foster, RI.*

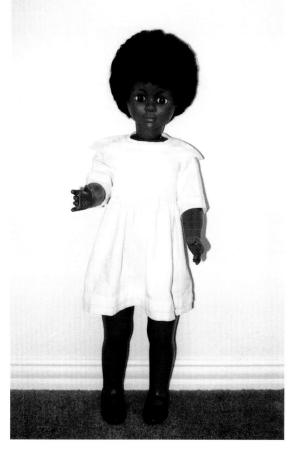

EEGEE

This Eegee girl, marked Eegee, has a light, almost porcelain look to her with a sweet face and rich blue eyes. In played with condition and redressed, she is worth about $30.

This 31 inch African American companion doll has authentic ethnic features, rather than being a darker version of a face mold meant for a white doll. She is valued at more than the white Eegee doll, because black dolls are harder to find - but less than the other black companion dolls in the chapter, because she appears to be more recent than the early 1960s. She is marked "Eegee Co. 31 E" and is valued at about $45. *Courtesy of Dian Hebert.*

ALLIED\EASTERN DOLL COMPANY

These companion dolls frequently marked with an "AE" have a sort of "pinched" look to their face. They are of medium quality and often found with a generous amount of hair. In good condition, they are worth about $40. Pigtailed Doll is marked AE 3651. *Courtesy of Diane Haddox*.

This radiant "AE" doll is just a dazzling example of a platinum blond. She is marked "AE 5651" on her neck. *Courtesy of Sherrie McCloughen.*
VALUE, BECAUSE OF THE UNUSUAL BEAUTY OF THIS DOLL, ABOUT $60

Doll with short blonde hair and cowboy boots. *Courtesy of Manuel and Mary Cotta.*

Striking African American version of the Allied Eastern doll, marked "AE 3651 / 33 ". Value: $60. *Courtesy of Mary Williams.*

I have seen this doll featured in "Identifying Your Dolls" type columns, and no one has been able to ascertain her manufacturer. She is marked "DE/ 32." With a face that appears identical to the Allied Eastern dolls, she, as an African American doll in wonderful condition, is valued at about $60. Thank you again to Debbie Garrett!

Another "DE/32" doll (left) stands next to a Uneeda African American Playpal lookalike, which is marked "U - 5" on her head. The Uneeda doll, another real beauty, is valued at about $60. *Courtesy of Debbie Garrett.*

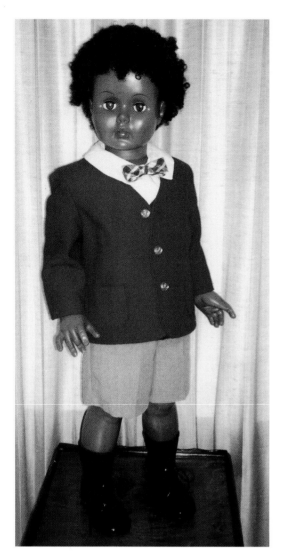

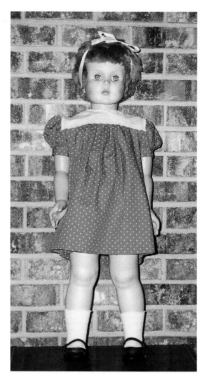

Allied Eastern Playpal lookalike marked "AE 5651 - 38." Valued at $40. *Courtesy of Mary Williams.*

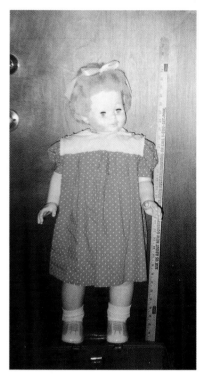

Eugene's "Baby Two Year Old" is 32 inches tall and is marked "Eugene Doll / 1978 / 53100." She is redressed, but wearing her original shoes. This is a more recent doll, but well liked by collectors of companion dolls. Value: $25. *Courtesy of Mary Williams.*

What a novelty! This Allied Eastern doll has been rewigged and redressed as a little boy! The end result is gives an impression of a little boy on his way to Sunday School, and is darling! "He" is marked "AE 5651 / 33." *Courtesy of Debbie Garrett.*

LARGER COMPANION DOLLS

This 32 inch Saucy Walker lookalike is unmarked and worth about $60. *Courtesy of Kathy Hosteller.*

Larger companion dolls (40" or taller) have higher values, even when they are unmarked, because they are harder to find, and collectors appreciate their uniqueness. These dolls were often used as mannequins in children's shops. This 40" unmarked companion doll is in wonderful condition and is valued between $100-125. *Courtesy Miss Nancy.*

SAYCO

27" unmarked companion doll very nicely redressed. In this very good condition she is worth $40. *Courtesy of Lani Spencer-McCloskey.*

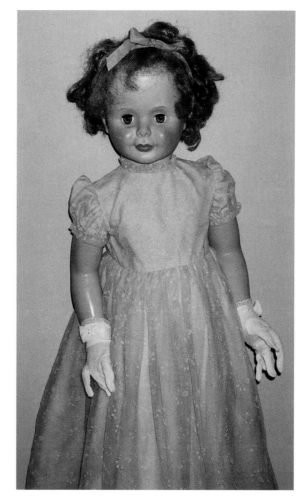

36 inch Playpal lookalike marked "Sayco" on both head and neck. Her value is about $40. *Courtesy of Mimi Smith and Debbie Garrett.*

The 35 inch companion doll manufactured by Sayco uses the Playpal limbs and has very pronounced lip color and very bright eyes when found in excellent condition. This doll is valued at about $40. *Courtesy of Lani Spencer-McCloskey.*

BUSTER BROWN STORE MANNEQUIN

Companion doll collectors very frequently love child - size store mannequins. They were not mass - produced in the way dolls were, and are, therefore much harder to come by. There is a certain allure to them, because as children, shopping with our Moms, these store mannequins were "untouchable." They looked like giant dolls or like frozen children, but were "roped off," so to speak, in the store window, the way artwork is roped off in a museum. As much as we wanted to, we never got to touch!! Just like clouds and your mother's Limoges - if you aren't allowed to touch it, it becomes ever so much more interesting!

AUCTION NOTE: Because of the beautiful clothing that was included, this Sayco Companion doll sold at auction for just under $90! Taking the time to dress up your companion dolls can appreciably add to their value. *Courtesy of Sheila Swierczewski.*

This mannequin is even more interesting because he has a name, and a very famous one at that - Buster Brown. Buster Brown was and is a well-known manufacturer of children's clothes and shoes. He is hard plastic, stands at 34 1/2 inches, and is attached to his stand with 2 screws. Manufactured by Old King Cole, Inc., he is marked "Buster Brown / Dress With Size 2 Garments." Value is about $250. *Courtesy of Lori Kelley.*